Schriever Air Force Base

A History in Pictures

For sale by the Superintendent of Documents, U.S. Government Publishing Office
Internet: bookstore.gpo.gov Phone: toll free (866) 512-1800; DC area (202) 512-1800
Fax: (202) 512-2104 Mail: Stop IDCC, Washington, DC 20402-0001

ISBN 978-0-16-093742-2

Acknowledgements

The preparation of this pictorial history of Schriever AFB was truly a group effort, involving many people and agencies within and outside the 50th Space Wing. Special thanks to the following people, without whose assistance this publication would have been much more difficult. Dr. Rick W. Sturdevant, Air Force Space Command History Office, provided guidance and editorial reviews and mentored me throughout the process of researching and preparing the book. Master Sergeant Scott D. Bailey, 310th Space Wing Historian, and Technical Sergeant Denise L. Flory, 302nd Airlift Wing Historian, provided much needed background information and photographs on the 310th Space Wing, Space Innovation Development Center, and USAF Warfare Center organizations and activities at Schriever AFB. Ms. Chris Blake , 50th Space Communications Squadron Multi-Media contractor, did the final layout and formatting and ensured the reprinted photographs met strict quality standards. The Space and Missile Center History Office staff provided photographs and key information on the early organizations supporting Air Force space activities. Major Michael Schriever was kind enough to loan several photos from family albums that provided insight into the life of his grandfather, General Bernard Schriever.

General Bernard A. Schriever, 1910 – 2005
"The father of Air Force missiles and space"

Table of Contents

The Airman's Creed

I am an American Airman.
I am a warrior.
I have answered my nation's call.

I am an American Airman.
My mission is to fly, fight, and win.
I am faithful to a proud heritage,
A tradition of honor,
And a legacy of valor.

I am an American Airman,
Guardian of freedom and justice,
My nation's sword and shield,
Its sentry and avenger.
I defend my country with my life.

I am an American Airman:
Wingman, leader, warrior.
I will never leave an Airman behind,
I will never falter,
And I will not fail.

Just Fence Me In!

The Ballad of Schriever AFB

Lyrics by Don Schaarschmidt

(To the tune *Don't Fence Me In*)

Oh give me land filled with sand
On the prairie side of town,
Just fence me in.

Let me drive Ninety-Four
Where the accidents abound,
Just fence me in.

Let me work through the day
Where the wind blows,
Trapped in an office
Without any windows,
Except on my computer
Where all my time goes,
Just fence me in.

I'll swipe my badge
Then I'll enter through the center,
Where security is high.

They'll check my bag
So no unpermitted camera,
Will see dishes in the sky.

I wanna walk on the lawn
With rabbits grazing,
Jog by the ponds
Those ducks are amazing,
I can't leave early
The Commander is gazing,
Just fence me in -
Just fence me in!

INTRODUCTION

Human desire to reach the stars is as old as civilization itself. Philosophers, scientists and engineers have looked to the heavens and considered ways to explore the galaxies. The Chinese developed gunpowder rockets by the first millennium A.D. and by 1045, gunpowder rockets were an important element of their military arsenal. According to legend, a minor Chinese official named Wan Hu even attempted, sometime during the 16th century, to launch himself to the moon using 47 rockets strapped to his chair. Centuries passed before humans developed the technologies to take the first real steps toward reaching space. In the early 20th century, pioneers such as Konstantin Tsiolkovsky, Hermann Oberth, Wernher von Braun, and Robert H. Goddard advanced the science of rocketry to new summits. Von Braun's rocketry work led to development of the German V 2, considered the foundation for some of today's modern rockets. World War II brought not only rockets, but jet engines, radar, and many other technological advances to the world. Tying those developing technologies together to foster exploration and exploitation of space would become the life's work of countless more scientists and engineers. In the late 1950s, the United States and the Soviet Union engaged in a "space race" to develop rockets capable of carrying artificial satellites into Earth orbit. The development of those rockets was also essential to the ongoing work in both nations to develop guided missiles capable of delivering nuclear weapons at intercontinental ranges.

In 1954, the U.S. Air Force established the Western Development Division (WDD). President Dwight Eisenhower announced the first U.S. scientific satellite program and WDD secretly initiated work on Weapon System 117L, the first Air Force satellite program. In 1956, the U.S. government awarded Lockheed Missiles and Space Company a contract for development of the country's first reconnaissance satellite, which eventually carried the top secret name "CORONA." On October 4, 1957, the Soviet Union launched Sputnik I, the world's first artificial satellite. The Soviets followed one month later with the launch of Sputnik II, which carried a dog named Laika (Barker), the first living creature placed into Earth orbit. The United States launched its first satellite, Explorer 1, on January 31, 1958, followed by Vanguard 1 on March 17, 1958.

CORONA, which first operated under the "Discoverer" scientific cover, originally was planned for launch from Cape Canaveral, Florida, and would be supported by satellite tracking facilities around the world. Some of those stations were already in place on Pacific islands to support missile tests. By 1958, the Air Force Ballistic Missile Division (renamed from the Western Development Division) determined that additional tracking stations were required and added Vandenberg Air Force Base and Point Mugu, California; Ka'ena Point, Hawaii; and Kodiak, Alaska, to the network. The launch site for Discoverer/CORONA also changed from Cape Canaveral to Vandenberg AFB. Discoverer I launched on February 28, 1959, and served as a test platform for guidance and booster systems. It transmitted telemetry intermittently until it de-orbited on March 17, 1959.

The Air Force quickly realized it needed an organization dedicated to on-orbit commanding and controlling of satellites. On April 6, 1959, the 6594th Test Wing activated at Lockheed's facility in Palo Alto, California, and the Air Force Satellite Control Network was born. At its activation, the 6594th included three operating locations: Edwards Air Force Base, California; Chiniak, Alaska; and Annette Island, Alaska. Later in the year, stations at Vandenberg AFB, California; Ka'ena Point, Hawaii; and New Boston, New Hampshire, joined the wing. In March 1960, the wing moved to Sunnyvale, California, and in June of that year, the new installation, designated as the Air Force Satellite Control Center, opened. Later, it would be redesignated Onizuka Air Force Station, in honor of Lieutenant Colonel Ellison Onizuka, an Air Force officer and NASA astronaut from California. The wing was redesignated the 6594th Aerospace Test Wing in 1961 and reassigned to the newly created Space Systems Division.

While the United States advanced its development of intercontinental ballistic missiles, it also continued work on its satellite programs, especially the CORONA reconnaissance satellite. On August 19, 1960, a specially equipped C-119 aircraft and its crew conducted the first aerial recovery of a film canister ejected from a CORONA satellite—Discoverer XIV. That satellite had acquired 3,000 feet of imagery covering more than 1.65 million square miles of Soviet territory. While the Discoverer program ended in 1962 after thirty-eight satellite launches, the CORONA program continued operations in secrecy for another ten years. Meanwhile, expansion and modernization of the satellite tracking stations continued as the United States began developing early warning, communications, and weather satellite systems. By the end of 1961, the network consisted of the main node at Sunnyvale and tracking stations at Vandenberg, Kodiak Island, Ka'ena Point, New Boston, Thule, Annette Island, Fort Greely, and by 1962, on the Seychelles island of Mahe in the Indian Ocean. By June 1962, the network was able to support multiple satellite operations. Network operators supported as many as ten programs, including their primary mission, CORONA.

The AFSCN gained responsibility for all ICBM test-flight and space tracking activities in 1963, and its organization and operations continued to expand. Recognizing the need for a new organizational construct to meet the growing network requirements, the Air Force established the Air Force Satellite Control Facility (AFSCF) as an organization in June 1965 and named Col. Walter H. Hedrick, Jr., as its commander. The AFSCF operated remote tracking stations around the world and conducted satellite telemetry, tracking, and commanding tasks now performed by the 50th Space Wing's space operations squadrons. The Guam Tracking Station opened in September 1965 as the first mobile facility, though permanent structures were soon constructed. The Los Angeles-based AFSCF was comprised of a support squadron and detachment at Sunnyvale, the tracking stations, and the 6594th Recovery Control Group, which was charged with the primary task of recovering CORONA satellite payloads. The group also conducted rescue missions in the Pacific and supported NASA's Apollo space launches. The 6594th Recovery Control Group inactivated in 1986 after recovering tens of thousands of film canisters ejected from once-classified satellites such as CORONA, Hexagon, and Gambit.

As the development of satellite technology advanced, new opportunities to field systems that enhanced military capability presented themselves. Sir Arthur C. Clarke had envisioned of the feasibility of manned, geostationary-orbiting communication satellites in 1945. A decade later, John R. Pierce, who had coined the word "transistor" and was working for Bell Telephone Laboratories, envisioned the utility of communications "repeaters" in space. On June 16, 1966, the Air Force launched the first cluster of Initial Defense Satellite Communications System satellites from the Eastern Space and Missile Complex (ESMC) at Cape Canaveral, Florida.

Meanwhile, the Department of Defense, had begun development of the Defense Meteorological Satellite Program, a sun-synchronous polar orbiting satellite using visual and infrared sensors to collect twice-daily weather imagery for use in military operational planning. The first Defense Meteorological Satellite Program launch occurred on May 23, 1962.

Concurrently, the military services were involved in the development of various pathfinder satellite-based navigation systems. The U.S. Navy successfully launched its first Transit navigation satellite in 1960 and, in 1967, the first TIMATION (Time Navigation) satellite. The U.S. Army developed the Sequential Correlation of Range (SECOR) satellites, while the U.S. Air Force began work on Program 621B. In 1968, the Department of Defense created the Navigation Satellite Executive Committee (NAVSEC) to coordinate the efforts of these organizations; a joint program office, established in 1973 under Air Force leadership, merged the Navy's TIMATION program and the Air Force's 621B program under the name NAVSTAR Global Positioning System (GPS). The Navy's TIMATION-IV, renamed as Navigation Technology Satellite (NTS)-2 launched on June 23, 1977, as the first demonstration satellite managed by the joint program office. The first Global Positioning System satellite, NAVSTAR-1, launched on February 22, 1978, with ten additional launches over the next seven years, culminating on October 9, 1985.

These new systems, both operational and planned, required improved and expanded facilities to manage their operations. The Sunnyvale, California site became the center of that initial expansion. By the late 1970s, however, the need for a new location for U.S. national satellite operations became evident. Growth in the Sunnyvale area presented Air Force leadership with a variety of concerns, not limited to the cost of utilities and earthquake hazards. Encroachment by business and housing developments had constrained the Sunnyvale Air Force Station, prohibiting the construction of additional facilities to support emerging and planned missions.

In 1979, the Secretary of Defense authorized the concept for a Consolidated Satellite Operations Center (CSOC) to become the primary AFSCN and satellite operations center and to support the Air Force's Shuttle Operations and Planning Complex (SOPC). Over the course of the next four years, Air Force and Department of Defense officials evaluated many sites with the potential to support the new CSOC. The availability of land, water, and inexpensive utilities were important considerations. Equally important were concerns about encroachment upon the installation by commercial or residential development. The selected site had to have sufficient resources to allow for expansion and new missions.

The Air Force Satellite Control Network continued its support for National Aeronautics and Space Administration (NASA) activities, especially in monitoring telemetry data for NASA launches in the Apollo program from the Sunnyvale location, while preparing to expand that support to include the soon to be operational Space Transportation System (STS), commonly known as the space shuttle. Beginning with the April 12, 1981 launch of STS-1, the AFSCN's support to NASA included downloading telemetry during launch, on-orbit, and re-entry operations of the space shuttle.

Following several years of evaluations involving nearly 100 potential sites, the Air Force recommended a previously unlisted option—a site approximately ten miles east of Peterson Air Force Base, which sat on the eastern boundary of Colorado Springs. This site, in unincorporated El Paso County, consisted principally of state-owned land, was rural, and lacked the encroachment issues associated with other evaluated sites. Its location provided sufficient security and offered advantageous "look angles" for the antenna systems that would be affiliated with the center's operational activity.

Following negotiations with the State of Colorado, the state granted the federal government a deed to approximately 640 acres of land. On May 17, 1983, contractors broke ground for what would become Falcon Air Force Station (AFS), named for the nearby unincorporated town north of the installation. For over two years, contractors worked to complete sufficient facilities to open the base, including headquarters and operations buildings, support facilities, and infrastructure.

On July 8, 1985 the 2nd Space Wing activated in a ceremony at Falcon AFS, even though the installation was not complete enough to allow the new wing to occupy the facilities. A ribbon-cutting ceremony on September 26, 1985, symbolized the activation of Falcon AFS. Within weeks, the initial cadre of the 2nd Space Wing, consisting of about 185 military and civilian employees moved into work centers and offices. Space Command activated the 2nd Satellite Control Squadron (later 2nd Space Operations Squadron) and its Vandenberg detachment to assume responsibility for Global Positioning System (GPS) on-orbit operations.

By the late 1980s, with much of the operations centers and communication links completed, the 2nd Space Wing began accepting missions previously conducted at California sites. By 1989, the 1st Satellite Control Squadron, later redesignated as the 1st Space Operations Squadron, had assumed responsibility for Defense Support Program (DSP) satellites and some aspects of Global Positioning System (GPS) operations, including satellite launch support. Between 1987 and 1993, Air Force Systems Command and Air Force Space Command officials coordinated efforts to transition mission activity from Sunnyvale AFS to Falcon AFS and the 2nd Space Wing.

On October 1, 1987, Onizuka Air Force Station (renamed from Sunnyvale AFS in honor of Lieutenant Colonel Ellison Onizuka, who lost his life in the space shuttle *Challenger* disaster of January 1986) transferred from Air Force Systems Command to Air Force Space Command and, ultimately, fell under the control of the 2nd Space Wing at Falcon AFS. Concurrent with that transfer, Air Force Space Command activated a number of organizations to assume satellite operations and network missions.

Leaps in space-related technologies added importance to Falcon's consolidated space operations and brought new missions and organizations to the station. Reflecting this growth, Air Force Space Command renamed the installation Falcon Air Force Base on June 13, 1988. Two days later, Lieutenant General Donald J. Kutyna hosted a ground-breaking ceremony for the Falcon AFB Space Forecast Center. And on June 17, 1988, Detachment 7, 2nd Satellite Tracking Group activated at Falcon AFB to operate the new Colorado Tracking Station of the Air Force Satellite Control Network.

As the development of the installation and the maturation of its units continued, 1988 became a banner year for the 2nd Space Wing and Falcon Air Force Base. As winter approached, the wing's satellite control crews continued to prepare for the transfer of satellite commanding missions. On November 29, 1988, the 1st Satellite Control Squadron accepted satellite command authority for the Defense Support Program. Then, on December 26, 1988, crews of the 1 SCS conducted their first Falcon AFB support of a Defense Meteorological Satellite Program satellite.

One year later, as primary responsibility for satellite missions began to transfer to the 2nd Space Wing's Falcon Air Force Base units, the wing opened the Consolidated Space Operations Center's Operations Command Center, transferring the function from the Air Force Systems Command's Space Systems Division in an April 3, 1989, ceremony. Initial operational testing and evaluations were conducted from February 15, 1989, until March 9, 1989. In October, Air Force Space Command responded to the needs of a maturing 2nd Space Wing by implementing organizational actions to increase the wing's self-sufficiency and efficiency. Some organizations were renamed to clarify their missions and others, like the 1002nd Civil Engineering Squadron, were activated to improve mission performance. By December 1989, Air Force Systems Command's Space Systems Division began the process of formally transferring mission control centers to Air Force Space Command and ultimately to the 2nd Space Wing user organizations.

The 1st, 2nd, and 3rd Satellite Control Squadrons were activated and transitioned satellite control and command responsibilities to their organizations in the first five to seven years after the base's activation. Responsibility for DSP, GPS, Defense Meteorological Satellite Program (DMSP), and wideband communications, such as the Defense Satellite Communications System (DSCS) satellite command and control operations, realigned under the 2nd Space Wing. Falcon AFB became the primary satellite command and control organization for the USAF and, also supported other national and military systems. The formal transition of primary operations from California to Falcon AFB was completed in August 1993, following successful operational testing of communication links between the ground stations and the installation, and between the Falcon and Sunnyvale communications nodes. A Falcon AFB ceremony on September 27, 1993, marked the formal turnover of the Consolidated Space Operations Center from Air Force Systems Command to Air Force Space Command.

In September 1990, the Joint National Test Facility (later renamed Joint National Integration Center) opened at Falcon, and a few years later, Air Force Space Command activated the Space Warfare Center (later renamed Space Innovation and Development Center) at Falcon AFB. These new organizations necessitated additions to the base's infrastructure.

Changing strategic priorities in the early 1990s led to a reduction of United States military organizations and personnel in Europe. To maintain the lineage and honors of distinguished units, the Air Force chose to inactivate Falcon's 2nd Space Wing and activate the 50th Tactical Fighter Wing, renamed 50th Space Wing, to assume responsibility for the satellite control and network operations missions at Falcon. The 50th activated at Falcon AFB on January 30, 1992, and absorbed the personnel, equipment, facilities, and functions of the inactivated 2nd Space Wing.

Throughout the 1990s, the installation and its organizations continued to develop and mature. Planning for new missions brought additional units to the 50th Space Wing. Air Force Space Command activated the 4th Strategic Reconnaissance Squadron, effective April 30, 1992, redesignated it the 4th Space Operations Squadron, and assigned it the mission of operating the yet-to-be-fielded Milstar satellite program. Air Force Space Command activated the Space Warfare Center on November 1, 1993, and charged that organization with incorporating space capability in major command, theater, and campaign war plans . . . simulation, modeling and war gaming activities, and to lead development efforts to improve combat applications of space systems. On September 4, 1997, the 310th Space Group (later 310th Space Wing) joined the growing list of organizations at Schriever AFB and became the first space group in the Air Force Reserve. The 310th grew over time and was redesignated the 310th Space Wing in March 2008. The Joint Forces Component Command (JFCC) for Integrated Missile Defense (JFCC-IMD) began operations at Schriever AFB in January 2005. Today, the base hosts twenty-one tenant organizations and activities and added another with the activation of the Joint Interagency Combined Space Operations Center (JICSpOC), which began operational experimentation and testing on October 1, 2015, while the 50th Space Wing operated or supported more than 175 satellites. The base's proximity to key space organizations, including Air Force Space Command headquarters, the Cheyenne Mountain Complex, and the growing space industry along Colorado's Front Range, were important factors in the base's growth over its first thirty years.

With a growing workforce, the 50th Space Wing moved to reduce the installation's reliance on nearby Peterson Air Force Base for support functions, beginning in the early 1990s. The 50th Mission Support Squadron (later Force Support Squadron) opened its own military and civilian personnel functions as well as Morale, Welfare, and Recreation (MWR) activities on September 13, 1992.

Over the next decades, Falcon continued growing as new missions necessitated increased land area for operations, support, and administrative facilities, as well as a buffer zone for security. In November 1993, the Air Staff proposed a land exchange with Colorado to obtain the desired properties. By February 1996, negotiations on land transfers with the State of Colorado, combined with purchases of privately owned parcels, resulted in acquisitions bringing the base's total area to nearly 4,000 acres. This provided both room for the base to expand and an adequate buffer against encroachment.

As the new millennium neared, the installation continued to grow. In 1997, Air Force Space Command activated the Space Battlelab at Falcon AFB to develop new and innovative ideas for applying space technology to combat operations. In 1998, the Air Force renamed the base to honor General Bernard A. Schriever, the man known as the "father of Air Force missiles and space." On June 5, 1998, the wing held a renaming ceremony in General Schriever's honor, marking a rare instance when an Air Force installation was named for a living person. Also, in 1998, construction began on new facilities to house missions and support operations being transferred from Onizuka Air Force Station as a result of the 1995 Defense Base Closure and Realignment Commission initiatives passed by Congress and signed into law by President William J. Clinton. Air Force Space Command ordered the activation, effective June 1, 2003, of the 50th Logistics Readiness Flight (previously designated the 50th Supply Squadron). The command followed that action with another unit activation, ordering the activation of the 50th Comptroller Flight (later elevated to squadron level), effective October 1, 2003, to provide financial services to Schriever AFB units and personnel.

Midway through the first decade of the 21st century, Schriever AFB hosted nearly sixty major and minor facilities and employed over 6,200 people. The base's continuing growth and importance prompted wing and command officials to begin preliminary planning to bring several hundred housing units and associated community support activities to the base. That effort came to fruition when, on May 16, 2008, base officials, community leaders, and contractor representatives broke ground on 242 enlisted and officer housing units at Schriever AFB. Construction was completed by late summer 2010. Today, 223 families call Schriever AFB home. By May 2014, Schriever AFB encompassed 3,840 acres holding sixty-three buildings (not including housing units) and reported a population of 7,647 personnel.

In a little more than three decades, Schriever has grown from humble beginnings as a "commuter base" to a full installation hosting many important missions and units from the U.S. Army, U.S. Navy, and U.S. Air Force, as well as joint organizations. The future looks bright for this base on the plains, because the conditions that made this site ideal for a new installation, still exist.

The history of Schriever Air Force Base (AFB), on the high plains of Colorado, may be brief in comparison to many other military reservations, but that history is filled with important accomplishments and community involvement.

Portrait of General Bernard A. Schriever

GENERAL BERNARD ADOLPH SCHRIEVER

General Bernard A. Schriever, born in Bremen, Germany on September 14, 1910, is considered the "father of Air Force missiles and space," having been an important contributor to America's development of ballistic missiles and space systems. Following World War II, the development of nuclear weapons and ballistic missiles had the most significant influence on the United States Air Force and the balance of world power.

The Schriever family immigrated to the United States in 1917 to join his father Adolph, whose ship had been seized in New York in 1916 and the crew interned. Following his father's release, the family moved to New Braunfels and onward to San Antonio, Texas, where Adolph found work in an engine plant. Adolph died in 1918 in an industrial accident, leaving his widow, who lacked skills and English fluency, as the sole supporter of herself, Bernard, and his younger brother, Gerhard. Following his father's death, Bennie and his brother lived in a foster home for eight months until their grandmother arrived from Germany to care for them. He became a naturalized citizen in 1923. After attending grade and high school in San Antonio, Texas, he attended and graduated from Texas A&M University with a bachelor of science degree in architectural engineering.

He entered the U.S. Army as a graduate of the Reserve Officer Training Course (ROTC) and received a reserve commission in the Field Artillery, but began flight training at Randolph Field in 1932, earning his wings and a commission in the Air Corps in 1933 at Kelly Field. He was assigned as a bomber pilot at March and Hamilton Fields, California, and received promotion to first lieutenant in June 1933. During the winter of 1934, he flew mail missions in antiquated and inadequately equipped plans that cost many of his comrades their lives. This experience underscored for him the consequences of technical inferiority and demonstrated the imperative to modernize and build up air power if the United States wanted to compete with other air forces.

Unable to secure a regular commission (instead of his current reserve commission), Schriever left the Air Corps in 1935 to fly as a commercial pilot. His application for a job with the airlines was not successful, so he took a job as a camp commander with the Civilian Conservation Corps in New Mexico. He left that position in October 1936, and returned to active duty with the Air Corps and in December 1936 he was assigned to Albrook Field, Panama. While assigned at Albrook Field, Schriever met his first wife, Dora Brett, the daughter of General George Brett. In August 1937, Schriever received his commercial pilot job with Northwest Airlines, and in January 1938, he married Dora Brett in the Washington, D.C. home of Brigadier General Henry H. "Hap" Arnold, a personal friend of the Bretts. Schriever returned to the Air Corps in 1938, receiving a regular commission as a second lieutenant instead of his pending promotion to captain in the reserves. He was assigned to the 7th Bombardment Group at Hamilton Field, California, as a B-18 instrument flying instructor.

In 1939, Schriever was reassigned to Wright Field, Ohio, where he later was one of six officers to attend the Air Corps Engineering School. After completing that course, the Army Air Forces (renamed from Air Corps in June 1941) sent him to Stanford University for a master's degree in aeronautical engineering. By June 1942, when he graduated from Stanford, Schriever had been promoted to major and was reassigned to the 19th Bombardment Group in the Southwest Pacific. Schriever participated in the Bismarck Archipelago, Leyte, Luzon, Papua, North Solomon, South Philippine, and Ryukyu campaigns, flying 33 combat missions in B-17s as the group's chief of maintenance before being removed from operations and moving to logistics and engineering. By the war's end, Schriever had been promoted to colonel and commanded the Advanced Headquarters, Far East Service Command, responsible for building and operating all maintenance, repair, and supply depots in the theater from bases in Hollandia, New Guinea, Leyte, Manila, and Okinawa.

His experience, education, and training, as well as his intellect and reputation, ensured that he remained involved in Air Force technical activities following the war. In 1946 he was assigned to the Pentagon to head the Scientific Liaison Branch under the Deputy Chief of Staff for Materiel. Schriever was entrusted with the task of maintaining close ties that had been forged during the war between the Army Air Forces (AAF) and leading scientists. Schriever worked closely with Dr. Theodore Von Karman, head of the Scientific Advisory Board (SAB). Von Karman introduced Colonel Schriever to many top scientists and helped him to reestablish the Air Force's connection to the scientific community that some believed had been neglected since General Arnold's retirement in February 1946.

Schriever soon found himself at the center of a small group of officers led by Major General Donald L. Putt, Air Staff Director of Research and Development, that campaigned for a separate research and development command and the establishment of a deputy chief of staff for development on the Air Staff. The group's efforts, aided by retired Lieutenant General James H. Doolittle, came to fruition and prevailed upon the Air Force to commission a study of research and development. The resulting report from the Ridenour Committee agreed with Schriever and his group that Air Force research and development would be improved with the creation of a research and development command.

After attending the National War College in 1950, Schriever was named to be Deputy Assistant for Evaluation under the new Deputy Chief of Staff for Development. Schriever's office was renamed the Development Planning Office in January 1951, and introduced development planning objectives (DPOs), an invaluable series of planning documents that matched long-range military requirements with ongoing research and development efforts. DPOs were generated for all major elements of air power—strategic and tactical warfare, airlift, air defense, and intelligence and reconnaissance. Schriever sought, through these DPOs, to push technology forward to meet future needs rather than pulling it to meet contemporary demands. In January 1951, he became the assistant for development planning. He was promoted to brigadier general in June 1953. While at the Pentagon, Schriever was instrumental in developing the B-58 Hustler and pushing for a low-altitude bombing capability.

General Schriever's work in R&D and acquisition continued and in June 1954 he was reassigned as the assistant to the commander of the Air Research and Development Command. One month later, he and a small group of officers headed to Los Angeles to organize and form the Western Development Division, predecessor to the Air Force's Ballistic Missiles Systems Division.

General Schriever made his most significant contributions in the development of the Air Force's intercontinental ballistic missile (ICBM) program. In May 1954, the Air Force assigned the Atlas missile top priority and directed that only technological concerns limit the speed of its development. Schriever became manager of the program at the

Western Development Division Headquarters.
Schriever's headquarters had once been a parochial school.
(Courtesy SMC History Office)

request of Trevor Gardner in mid-1954 after it had suffered many challenges, including low prioritization and indecision. General Schriever required that he be given broad authority to complete his task. In August 1954, Schriever assumed command of the Western Development Division (WDD) and also served as Assistant to the Commander, ARDC. This second title gave the general a path around much of the bureaucracy.

Schriever's WDD gained the Titan project as a parallel to the Atlas missile program as well as the Thor, Navaho, and Snark missile programs. And, while the Navaho did not result in an operational missile, its builder, North American, went on to develop the rocket engines for military missiles and space launch vehicles. By 1956, Schriever had succeeded in having WDD named as the responsible agency for the development of satellite systems. In early 1957, he gave a controversial speech in San Diego calling for the United States to establish space superiority. Less than eight months later, the Soviet Union launched Sputnik I, the world's first man-made satellite. The National Aeronautics and Space Administration, established in October 1958, relied heavily on the Air Force's Thor, Atlas, and Titan missiles as boosters for its space operations. Then, on February 28, 1959, the Air Force launched Discoverer I from Vandenberg AFB, California, and laid the foundation for the era of satellite-based reconnaissance with the launch of Discoverer 14, on August 18, 1960.

In 1959, General Schriever received promotion to lieutenant general and assumed command of the Air Research and Development Command. There, he implemented concurrency in systems development and acquisition. Concurrency, a concept in which the Air Force initiated the conceptual phase of a new system then handed the acquisition to systems centers that refined the system during its operational phase, had served WDD well in the development of its assigned programs. This approach compressed acquisition time and got new systems into the hands of operators more quickly.

As head of ARDC, Schriever pushed to have all responsibility for weapon systems acquisition transferred to his command from the Air Materiel Command (AMC). He had faced this challenge as commander of the WDD, as well. Despite some success, General Samuel E. Anderson, then commander of the AMC, objected to further integrations and proposed a reorganization that would put ARDC under AMC, a proposal to which Schriever objected. The generals each stuck to their positions, and no resolution occurred for the next two years. During the Kennedy administration, General Thomas D. White, Air Force Chief of Staff, sided with Schriever, and the reorganization occurred on Schriever's terms. ARDC became the Air Force Systems Command with General Schriever, who received his fourth

General Schriever discusses the mobile Minuteman project with Boeing president William Allen and Major General Osmond Ritland.

star, at the helm; AMC became the Air Force Logistics Command. Schriever established the Space Systems Division at Los Angeles in April 1961. He remained in command of the Air Force Systems Command until his retirement on August 31, 1966.

In June 1998, Falcon AFB in Colorado was renamed in honor of General Schriever, recognizing his significant contributions to Air Force missile and space programs.

General Bernard Adolph Schriever died on June 20, 2005 at his home in Washington, D.C.

General Schriever took great personal pride that, in his many years of dealing with industry, not one official protest was lodged concerning irregularity in selecting contractors for the enormously costly ballistic missile programs. Schriever himself often was singled out by lobbyists seeking favorable treatment for their clients. All failed. Schriever built a record of unquestionable integrity.

Elizabeth Schriever with sons Bernard (right) and Gerhard (center). Unknown date. Following her husband's death in an industrial accident, Elizabeth, lacking job skills and English fluency, became the family's sole source of support. Eventually, she found work as housekeeper for a wealthy San Antonio, Texas family and sold homemade sandwiches and cold beverages to golfers at the Breckinridge municipal golf course, which bordered on her small house—built for her by her employer. *(Photo courtesy Schriever family)*

In high school, Bennie earned money doing chores and caddying at the local golf course. He became an exceptional golfer and won the San Antonio, Texas, Junior Amateur Championship in 1931. He was featured in Ripley's Believe It Or Not for thrice hitting a drive more than 300 yards to the same green and one-putting for eagle. Following his graduation from Texas A&M University, he faced the choice of becoming a professional golfer or, because he had been an ROTC cadet, accepting a reserve commission in the U.S. Army field artillery. He chose the latter, but soon became interested in flying and transferred to the U.S. Army Air Corps.
(Photo courtesy Schriever family)

Aviation Cadet Bennie Schriever, shown here in his flying gear, attended initial flying training at Randolph Field (now a component of Joint Base San Antonio) near San Antonio, Texas, and then underwent advanced flying training at Kelly Field (later Kelly AFB and now Kelly Field Annex), in San Antonio, Texas. Despite his love of flying, his aptitude in mechanics and engineering would play a much larger role in his career than he probably imagined as a young aviator. *(Photo courtesy Schriever family)*

Following the completion of training, Bennie Schriever received his commission as a second lieutenant and transferred to March Field (now March Air Reserve Base), near Riverside, California. At March Field, Lieutenant Schriever joined the 7th Bombardment Group as a pilot under the command of Major Carl A. Spaatz. The 7th Bombardment Group operated the Curtiss B-2 Condor and Keystone B-4 Panther bombers. He moved with the 7th Bombardment Group to Hamilton Field (formerly Marin County Air Field). Schriever then flew airmail from Salt Lake City, Utah, to Cheyenne, Wyoming, in ill-equipped O-38 (photo) and B-4 aircraft. The experience in the "airmail fiasco" instilled in the young officer an appreciation for the price to be paid for inferior or inadequate technology. *(Photo courtesy Schriever family)*

Schriever left the Air Corps in 1935 to pursue a career as a commercial pilot. When he did not receive a position with an airline, however, he found employment as the commander of a Civilian Conservation Corps encampment in New Mexico. That job ended in 1936, and he received a regular Air Corps commission, being stationed at Albrook Field, Panama, with the 19th Wing. There he served as a P-12 pilot and assistant to Brigadier General George H., Brett. He married General Brett's eldest daughter, Dora, in January 1937 in the Washington, D.C., home of Brigadier General Henry H. Arnold. This P-12 is painted in the colors for the 6th Pursuit Squadron, 18th Pursuit Group, Wheeler Field, Hawaii.

This B-18, Bolo, aircraft is similar to that flown by Lieutenant Schriever when he returned to active duty in 1938 following a brief tenure as a commercial pilot with Northwest Airlines. As war clouds gathered over the Pacific, Schriever moved to Wright Field, Ohio, as a test pilot. While there, he attended the Air Corps engineering school. The Air Corps then sent him to Stanford University to complete a master's degree in aeronautical engineering. The Army Air Forces denied his request for assignment to a combat unit following the Pearl Harbor attack, requiring him to remain in California to finish his graduate work.

By July 1942, Major Schriever had his combat assignment and joined the 19th Bombardment Group in Australia. Despite his prowess as a bomber pilot—he flew thirty-eight combat missions in B-17s, B-25s, and C-47s—Schriever's commanders believed his skills and knowledge in engineering made him more valuable as the group's chief of maintenance. When the 19th was ordered back to the states, General George C. Kenney held Schriever in the Pacific, making him chief of the Maintenance and Engineering Division, 5th Air Force Service Command, in January 1943. His responsibilities and rank rose quickly and he assumed command of the advance headquarters, Far East Air Service Command responsible for maintenance for the 5th, 7th, and 13th Air Forces. He was promoted to colonel in December 1943. *(Photos courtesy Schriever family)*

Lieutenant General (Retired) Jimmy Doolittle was a proponent of Schriever's campaign, with a small cadre of other officers led by Major General Donald L. Putt, to establish a deputy chief of staff for development on the air staff. The Ridenour Committee later reported the results of their study, which concluded that Air Force research and development would be improved with the creation of a research and development command. The Air Force established Research and Development Command (later Air Research and Development Command) in 1950. ARDC was redesignated Air Force Systems Command on April 1, 1961. It inactivated on July 1, 1992.

Trevor Gardner congratulates Schriever on his promotion to major general in December 1955, after he assumed command of the Western Development Division the previous year. General Schriever also served as Assistant to the Commander, Air Research and Development Command. As the Assistant to the Commander, ARDC, Schriever was able to circumvent much of the organizational bureaucracy in managing the weapons systems he was charged with developing. As the Commander, Western Development Division, Schriever advocated for the development of a second ICBM for the Air Force. That proposal became the Titan ICBM.

Also in 1956, General Schriever proposed, then oversaw, deployment of the Thor Intermediate Range Ballistic Missile in England. Thor (right) became the Air Force's first ballistic missile system on alert during the Cold War. He also proposed and oversaw development of an upper-stage booster for the Atlas missile to support the Air Force's developing satellite program. That booster, the Agena (left), became the workhorse of the Air Force satellite program and carried many of the early satellites, including CORONA and MIDAS intelligence satellites, into high orbits around Earth. The WDD also gained responsibility for the Navaho and Snark missile programs.

In 1957, Western Development Division became the Ballistic Missile Division, and General Schriever remained its commander. Citing advancements in rocket propellants, casings, and vector controls, Schriever proposed the Air Force deploy a small, solid-fueled ICBM in large quantities. That missile, originally designated Weapon System Q, became the Minuteman. The program included early designs for a rail mobile version that would be less susceptible to Soviet attack. Its development also led to the development of submarine launched ballistic missiles. In February 1957, Schriever gave a speech in San Diego, California, where he called for the United States to obtain and maintain space superiority. Although admonished by Air Force senior leaders for being so outspoken, Schriever gained popular plaudits. On April 1, *Time* magazine put him on the cover as "Missileman Schriever".

In 1959, Schriever was promoted to lieutenant general and assumed command of Air Research and Development Command. He proposed development of a reusable space vehicle. The Aerothermodynamic/Elastic Structural Systems Environmental Test (ASSET) was a reusable spacecraft developed by ARDC (later renamed Air Force Systems Command). ASSET carried an instrumentation package that recorded the effects of reentry and flat gliding on space vehicles. The ASSET missions provided important information to engineers and scientists who were developing NASA's space shuttle.

A RAND report in 1946 cited the many effects that could be delivered from space—communications, weather, reconnaissance and intelligence, among those. By 1956, the Western Development Division had been given the responsibility for developing the U.S. military's first space systems, as well as its missile programs. While missile systems remained the priority, work began on Weapon System 117L, which by 1959 had evolved into three separate programs--Discoverer, the Missile Defense Alarm System (MIDAS), and the Satellite and Missile Observation System (SAMOS).

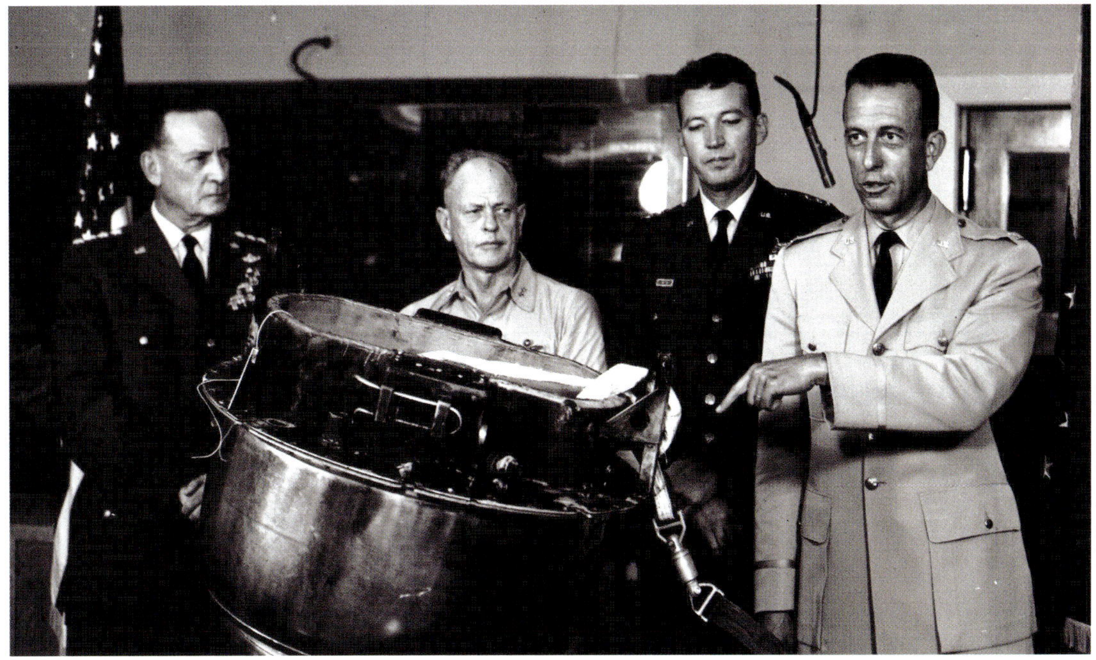

Discoverer I was the world's first polar orbiting satellite. The first of its series launched on February 28, 1959, on a Thor/Agena booster. The program under that name continued until February 1962, then fell under the secret code name CORONA and continued for another decade. The film canister payloads from Discoverer and CORONA missions were recovered by aircraft as they deorbited. In this photo, Lieutenant General Schriever (center right), Major General Osmond J. Ritland (center left), General Thomas D. White, Air Force Chief of Staff, and Colonel Lee Battle, CORONA Program Manager, discuss the recovery of the Discoverer XIII capsule—the first object recovered from space. The capsule carried no film, only an instrumentation package.

On April 1, 1961, under a reorganization of research, development, and acquisition functions, Air Research and Development Command became Air Force Systems Command. General Schriever received his fourth star and stayed in command of the newly designated organization, remaining its commander until his retirement on August 31, 1966.

Jacob Neufeld summarized General Schriever's career. "He will be best remembered as the architect of the Air Force's missile and space programs, but his influence went far beyond that. He also introduced systems management and operations research; project management; and systems engineering. Thus, he achieved the mission he had vowed to carry out when he flew the airmail in 1934, ensuring America's aerospace superiority. And he far exceeded his assignment from Hap Arnold in 1945, "to maintain the close ties between the Air Force and the scientific community." So successfully did Schriever merge science and engineering with military procedures that he created a methodology that became the standard throughout the Defense Department for many years."

Schriever remained active in government following his retirement, serving on numerous advisory boards and as a consultant. He remained a tireless advocate of advanced technology research. Before retiring, General Schriever told a meeting of the Arnold Air Society: "The world has an ample supply of people who can always come up with a dozen good reasons why a new idea will not work and should not be tried, but the people who produce progress are a breed apart. They have the imagination, the courage, and the persistence to find solutions."

General Schriever died at his home in Washington, D.C., on June 20, 2005.

USAF SATELLITE OPERATIONS, 1960-1980

General Schriever performed an important role in establishing the organizational structures that would develop and build the initial U.S. space capabilities, including satellites and missile systems, but also the infrastructure to support those systems.

The Western Development Division, created in 1954, by 1956, had been given responsibility for the development of satellite systems as well as intercontinental ballistic missiles. President Dwight Eisenhower announced the first U.S. scientific satellite program and the Air Research and Development Command transferred development of Weapon System 117L, the first Air Force satellite program, from the Wright Air Development Center to WDD on October 10, 1955. Weapon System 117L included three separate programs, the *Discoverer* reconnaissance satellite, the Missile Defense Alarm System (MIDAS), and the Satellite and Missile Observation System (SAMOS). In 1956, the U.S. government awarded Lockheed Missiles and Space Company a contract for development of the country's first reconnaissance satellite. The missile program had the highest priority, however, and work on satellite systems was often underfunded.

On October 4, 1957, the Soviets launched Sputnik I, the world's first man-made satellite. The Soviets followed that with the successful launch of the first biological specimen, a dog named Laika, into orbit on November 3, 1957. The Sputnik launches resulted in improved funding for the WDD's satellite programs as some military and government officials feared a Soviet advantage in the development of satellite systems.

Following these Soviet "firsts," the United States made quick headway, despite some early setbacks. On December 6, 1957, the U.S. attempt to launch the Naval Research Laboratory's TV3 satellite using a Vanguard rocket failed two seconds into the launch. The rocket fell back onto the launch pad rupturing the fuel tanks and causing an explosion. On January 31, 1958, the United States responded with the launch of Explorer I. It carried a cosmic ray detection experiment developed by Dr. James Van Allen, a physicist at the University of Iowa. The satellite detected fewer cosmic rays in its orbit, which ranged from 220 miles to 1,563 miles above the Earth, than Van Allen had anticipated. Van Allen theorized that the satellite's instrumentation had been saturated by passing through a belt of very strong radiation trapped in Earth's magnetic field. The next successful U.S. satellite launch, on March 17, 1958, carried Explorer III aloft using a Vanguard rocket. That satellite confirmed Van Allen's theory with the discovery of the Van Allen radiation belts. The intermediate Explorer launch, Explorer II, was terminated when the Jupiter C rocket's fourth-state failed to ignite.

President Eisenhower signed legislation formally establishing the National Aeronautics and Space Administration effective on October 1, 1958, to oversee the U.S. civil space program. NASA's first launch, Pioneer 1, occurred on October 11, 1958. Then, on December 6, 1958, NASA launched Pioneer 3, the first U.S. satellite to achieve an orbit of 63,580 miles. The Air Force, on December 18, 1958 placed a communications relay satellite, Project Score, into orbit aboard an Atlas booster. President Eisenhower's 1958 Christmas message was beamed from the satellite and the president's voice became the first beamed from space.

In its November 1958 Military Space Program Plan, the Department of Defense identified five primary functions for satellite systems. These included navigation, meteorology, communications, missile detection and space defense, and earth observation. In the spring of 1959, the Air Force added eight additional tasks for space assets including maintenance and supply of space vehicles, a manned moon station, bombardment satellites, and a manned detection, warning, and reconnaissance space vehicle.

Under General Schriever's direction, and benefitting from his talented team of managers, contractors, engineers, and scientists, the United States fielded its first- and second-generation intercontinental ballistic missiles and, on February 28, 1959, launched Discoverer I, the pathfinder satellite for a new space-based reconnaissance program. Discoverer I served as a test platform for guidance and booster systems. It transmitted telemetry intermittently until it deorbited on March 17, 1959. The Discoverer Program ultimately included thirty-eight satellite launches before it formally ended in February 1962. Thereafter, the reconnaissance program begun under Discoverer continued, under the top secret code name CORONA, for another decade.

Satellites on orbit require ground stations for tracking, telemetry, and ultimately commanding. The creation of a network of such stations to conduct these satellite support tasks would take a number of years to complete. By 1958, the Ballistic Missile Division (renamed from the Western Development Division) determined that tracking stations in addition to those supporting missile tests in the Western Range were required. Beginning with assets in use at the Western Range, the Air Force began establishing satellite tracking stations around the globe and moved to add Vandenberg AFB and Point Mugu, California; Ka'ena Point, Hawaii; and Kodiak, Alaska, to the network.

The Air Force quickly realized it needed an organization dedicated to on-orbit commanding and controlling of satellites. On April 6, 1959, the 6594th Test Wing activated at Lockheed's facility in Palo Alto, California, and the Air Force Satellite Control Network was born. At its activation, the 6594th included three operating locations--Edwards Air Force Base, California; Chiniak, Alaska; and Annette Island, Alaska. Later in the year, stations at Vandenberg AFB, California; Ka'ena Point, Hawaii; and New Boston, New Hampshire, joined the wing. In March 1960, the wing moved to Sunnyvale, California, and in June of that year, the installation, designated Air Force Satellite Control Center, opened. Later, the installation would be renamed Onizuka Air Force Station. The wing was redesignated 6594th Aerospace Test Wing in 1961 and reassigned to the newly created Space Systems Division.

Meanwhile, expansion and modernization of the satellite tracking stations continued as the United States began developing communications and weather satellite systems. By the end of 1961, the network consisted of the main node at Sunnyvale and tracking stations at Vandenberg, Kodiak Island, Ka'ena Point, New Boston, Thule, Annette Island, Fort Greely, and by 1962, on the Seychelles island of Mahe in the Indian Ocean. By June 1962, the network was able to support multiple satellite operations. Network operators supported as many as ten programs, including their primary mission, CORONA.

The AFSCN gained responsibility for all ICBM testing and space tracking activities in 1963, and its organization and operations continued to expand. Recognizing the need for a new organizational construct to meet the growing requirements of the network, the Air Force established the Air Force Satellite Control Facility as an organization in June 1965. The AFSCF operated remote tracking stations around the world and conducted satellite telemetry, tracking, and commanding tasks now performed by the 50th Space Wing's space operations squadrons.

The Guam Tracking Station opened in September 1965 as the first mobile facility, though permanent structures were soon constructed. The Los Angeles-based AFSCF was comprised of a support squadron and detachment at Sunnyvale, the tracking stations, and the 6594th Recovery Control Group, which was charged with the primary task of recovering CORONA satellite payloads. The group also conducted rescue missions in the Pacific and supported NASA's Apollo space launches. The 6594th Recovery Control Group inactivated in 1986 after recovering tens of thousands of film canisters ejected from once-classified satellites, such as CORONA, Hexagon and Gambit.

The network also continued its support for NASA activities, especially for the Apollo lunar missions and other experimental programs. The AFSCN would later support NASA during operations of its planned Space Transportation System. Supporting these new systems required improved and expanded facilities. The Sunnyvale, California, site had served as the center of that expansion.

As the development of satellite technology advanced, new opportunities to field systems that enhanced military capability presented themselves. Sir Arthur C. Clarke wrote of the feasibility of manned, geostationary-orbiting communication satellites in 1945. A RAND report in 1946 also proposed effects that could be delivered from satellite platforms, including reconnaissance and intelligence, weather, and communications. Less than a decade later, John R. Pierce, working for Bell Telephone Laboratories, wrote of the utility of communications "repeaters" in space. On June 16, 1966, the Air Force launched the first initial Defense Satellite Communications System satellites from the Eastern Space and Missile Complex at Cape Canaveral, Florida.

Meanwhile, the Department of Defense, by the early-1960s and following the success of the first Television Infra-Red Observation Satellite (TIROS) meteorological satellites, had begun development of the Defense Meteorological Satellite Program, a sun-synchronous orbiting satellite using visual and infrared sensors to collect twice daily weather imagery for use in military operational planning. Concurrently, the military services were involved in the development of various pathfinder satellite-based navigation systems. The U.S. Navy successfully launched its first Transit navigation satellite in 1960 and, in 1967, the first of its TIMATION (Time Navigation) satellites. The U.S. Army developed the Sequential Correlation of Range (SECOR) satellites, while the U.S. Air Force began work on Program 621B. In 1968, the Department of Defense created the Navigation Satellite Executive Committee to coordinate the efforts of these organizations. A joint program office, established in 1973 under Air Force leadership, merged the Navy's TIMATION program and the Air Force's Program 621B under the name NAVSTAR Global Positioning System. The Navy's TIMATION-IV, renamed as Navigation Technology Satellite-2, launched June 23, 1977, as the first NAVSTAR GPS demonstration satellite managed by the joint program office. The first GPS satellite, NAVSTAR-1, launched February 22, 1978, with ten additional satellite launches during the next seven years, culminating October 9, 1985.

From the MIDAS program that had been a component of WS 117L came the Defense Support Program ballistic missile warning satellite. First launched in November 1970 from the Eastern Space and Missile Complex, DSP became the foundational system for missile warning. With twenty-three launches over the next twenty-seven years, the satellite evolved through five generations. The first block launched from 1970 to 1973. The Air Force launched three Block II satellites from 1975-1977 and followed with the Block III, Multi-Orbit Satellite Performance Improvement Modification on four satellites between 1979 and 1984. Block IV, an upgrade to the Block II platform, occurred from 1984 to 1987 and was carried aboard two vehicles. The fifth generation of DSP satellites launched from 1989 until 2007.

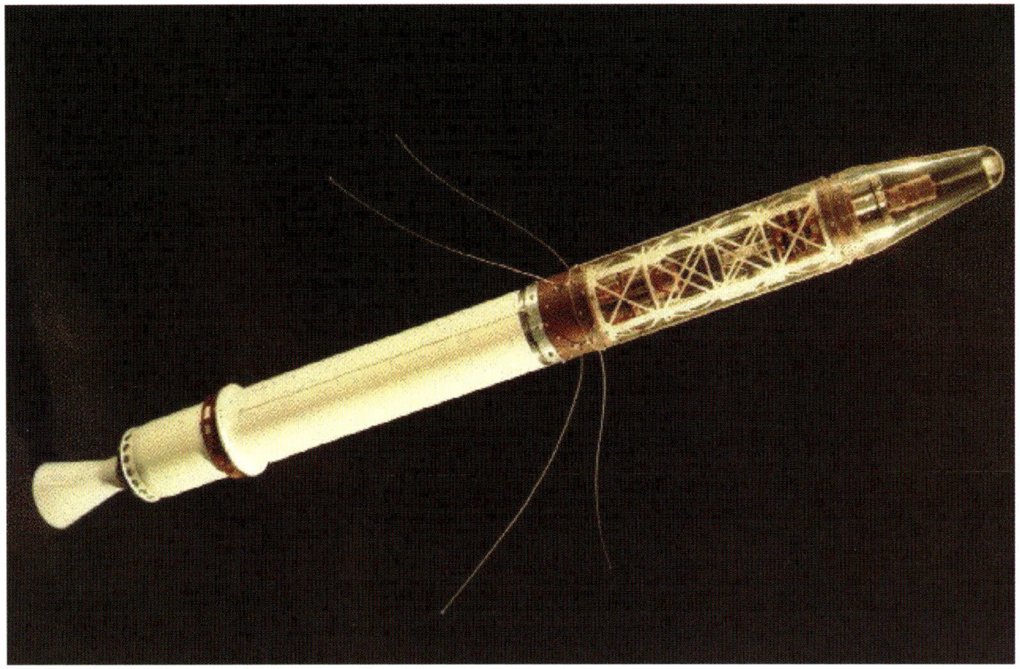

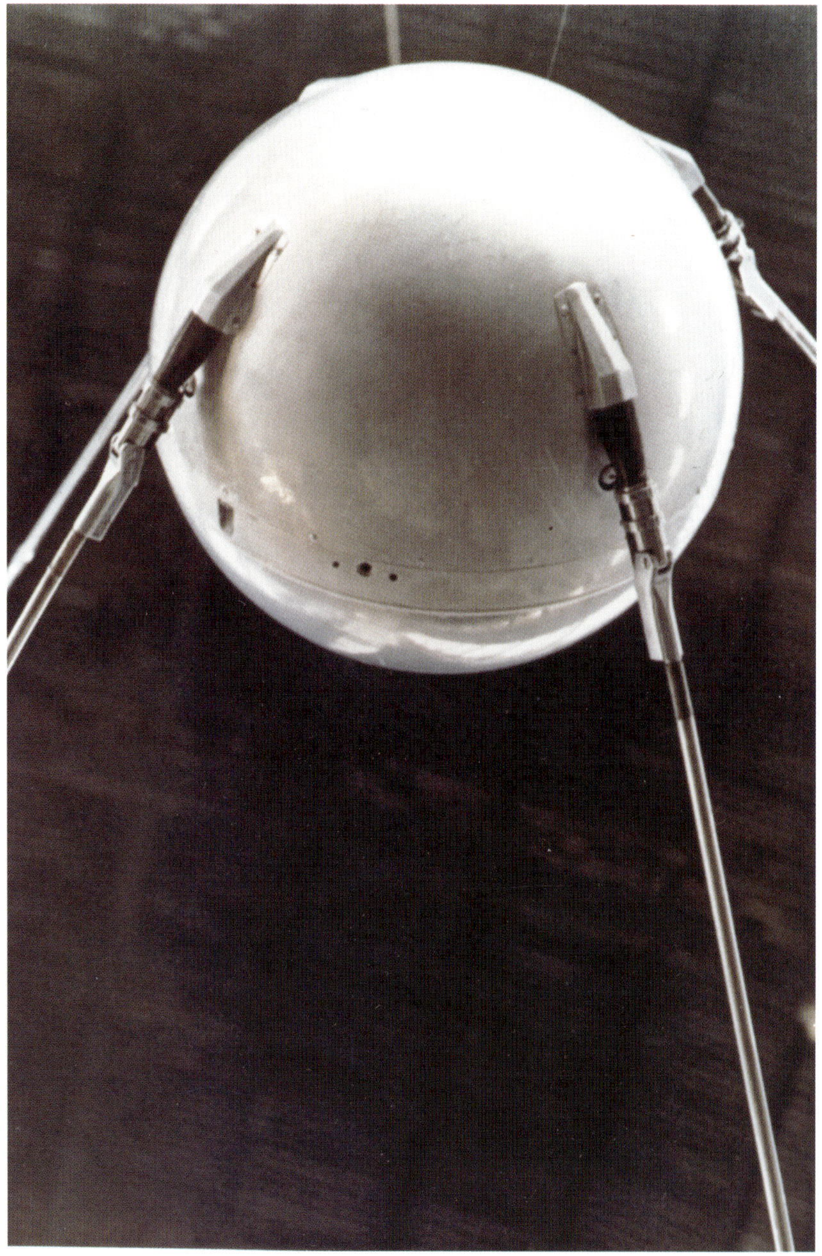

While tasked, in 1956, to develop U.S. military satellites in conjunction with their work on ICBMs, WDD (later renamed Ballistic Missile Division) satellite programs were underfunded compared to the missile programs that held highest priority. On October 4, 1957, the Soviet Union launched Sputnik I. The U.S. responded by hastening the production and launch of Explorer I on January 31, 1958. By then, the Soviet Union had already launched its second satellite, Sputnik II, carrying the dog Laika, and claimed first place in orbiting a biological specimen. Despite these "firsts" for the Soviets, General Schriever believed the United States held a technological advantage over the Soviets.

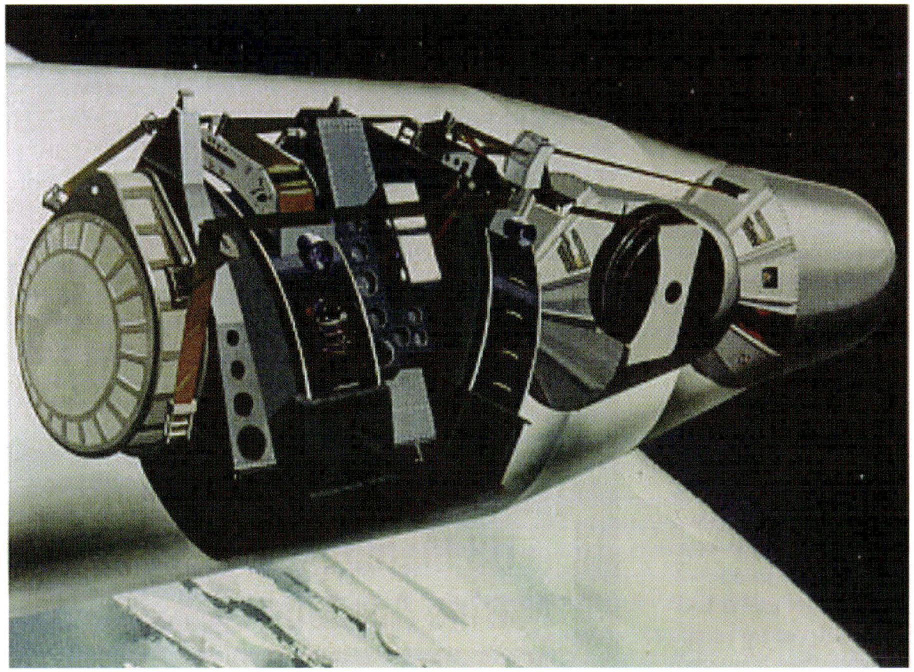

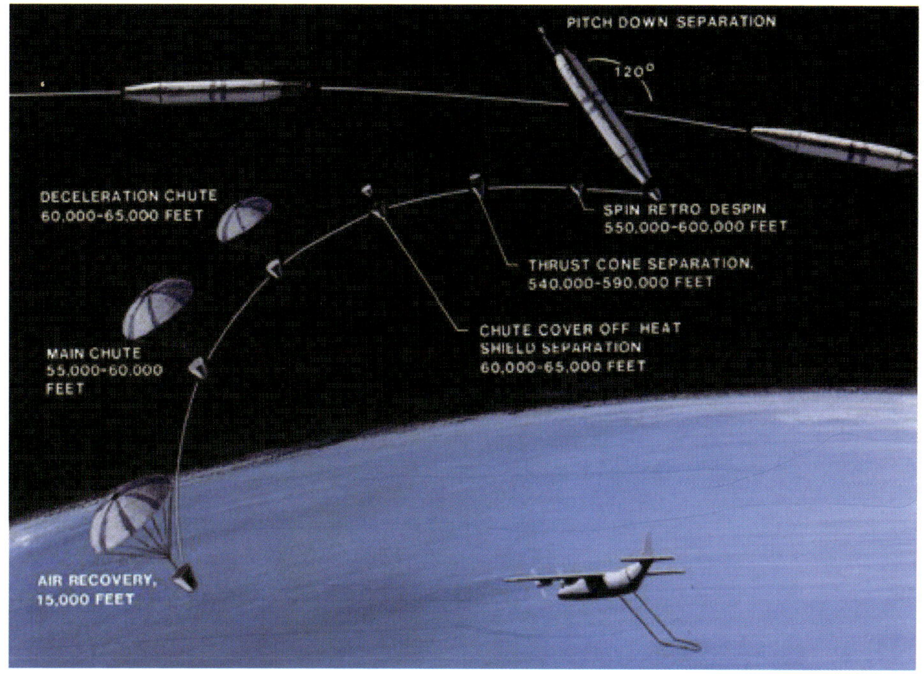

In February 1959, the United States Air Force launched Discoverer I, a pathfinder to test the performance capabilities of the propulsion and guidance systems of the program's satellite. The 618.0 kg satellite launched from Vandenberg AFB aboard a Thor booster. Discoverer I was 5.73 meters long and 1.52 meters in diameter. It consisted of an Agena upper stage capped by a conical nose cone comprised of a magnesium casing. The payload, consisting of communications and telemetry equipment was housed in the nose cone. This developmental satellite carried no camera or film payload. Discoverer I was the world's first polar orbiting satellite and continued to transmit telemetry data until it deorbited in March 1959. Discoverer XIV provided the first photographs of the Soviet land mass in August 1960. As the Agena A passed over Alaska on its seventeenth orbit, it ejected the film capsule. After the capsule reentered the atmosphere, it deployed a parachute that was spotted by the crew of a C-119J of the 6593rd Test Squadron (Special). The crew successfully snagged the parachute on the third pass and reeled the film capsule aboard. This was the first aerial recovery of an object returning from orbit and the first aerial recovery of film from an orbiting satellite.

By 1959, the Air Force realized it needed an organization specifically to conduct its satellite tracking and commanding functions. The Air Force organized the 6594th Aerospace Test Wing at the Lockheed Missile and Space Facility at Palo Alto, California, and assigned the organization to the Air Research and Development Command (later redesignated Air Force Systems Command). The unit was assigned to the Air Force Ballistic Missile Division of Air Research and Development Command on November 16, 1959. This new organization initially consisted of five squadrons. The 6595th Test Squadron joined the wing on February 20, 1961, and Detachment 1 at Donnelly Radio Relay Site in Delta Junction, Alaska, joined the wing on April 15, 1962.

The 6594th Aerospace Test Wing operated only briefly from the Palo Alto facility. By 1961, a new control center had been constructed at the Air Force Satellite Test Center (later renamed Onizuka Air Force Base) near Sunnyvale, California. As military space operations grew, so did the Satellite Test Center. The Sunnyvale complex would serve Air Force and national space activities for the next fifty years under the management of a variety of military organizations. The 6594th Aerospace Test Wing included tracking stations at Ka'ena Point, Hawaii (1959); Vandenberg AFB, California (1959); New Boston, New Hawmpshire (1959); Kodiak, Alaska (1959); Seychelles, British Indian Ocean Territory (1962); and Thule Air Base, Greenland (1962). The station at Annette Island, Alaska, which became operational by November 1958, closed in 1963. Its responsibilities were assumed by the Kodiak Island station. In 1965, the Air Force Satellite Control Facility, a named organization, replaced the 6594th Aerospace Test Wing as the primary network organization.

Before the recovery of the Discoverer XIV film capsule, the Television Infra-Red Observation Satellite (TIROS), designed by the Radio Corporation of America (RCA) launched into orbit on April 1, 1960. TIROS-1, the world's first weather observation satellite, took images of the Earth's cloud cover from an orbit of 450 miles. Although it operated for only seventy-eight days, the satellite demonstrated the feasibility of monitoring Earth's cloud cover and weather patterns from orbit. TIROS provided the first accurate weather forecasts from space-based data. In 1962, the USAF orbited the first of the Defense Meteorological Satellite Program (DMSP) satellites as Program 35, then later re-designated as the Data Acquisition and Processing Program. Initially, DMSP was highly classified and run by the National Reconnaissance Office, in support of the CORONA program and its first reconnaissance satellites.

In 1965, the Air Force added the Guam Tracking Station as Operating Location 10, Air Force Satellite Control Facility, opening the network's first mobile facility. Work quickly began, however, on the construction of permanent facilities. The tracking station's staffing stood initially at fifty-six persons, most of whom were civilian contractors of the Philco company. Major Steven A. King, station commander, described the arrival of Philco family members in his first monthly report:

"The wives of OL 10 personnel are beginning to arrive on Guam. To date, eleven Philco families are living on the Island. All seem affected to some degree by the substandard housing that exists on Guam, by the high cost of living, and by the oppressive heat and humidity. A period of at least one month seems necessary to become used to the tropical conditions on Guam. Any wives who are the least squeamish about bugs will probably require a longer period."

The Air Force was not the only military department developing satellite technology. In 1964, Naval Research Laboratory engineers, led by Mr. Roger Easton, conceived the Time/Navigation (TIMATION) program, building upon the U.S. Navy Transit satellite. The U.S. Army, meanwhile, was working on its Sequential Correlation of Range (SECOR) satellite and the Air Force began development of Program 621B. The U.S. Navy launched TIMATION 1 in 1967. In 1968, the Department of Defense created the Navigation Satellite Executive Committee to coordinate the efforts of the three services. Then, in 1973, the creation of a joint program office under the leadership of the Air Force merged the TIMATION and 621B programs, creating the NAVSTAR program. The Navy's next two TIMATION satellites were redesignated Navigation Technology Satellite (NTS) 1 and 2, with NTS-2, launched on June 23, 1977, the first NAVSTAR Global Positioning System (GPS) demonstration satellite managed by the joint program office. NAVSTAR -1, the first GPS experimental Block I satellite, launched on February 22, 1978, with ten additional satellite launches during the next seven years, culminating on October 9, 1985.

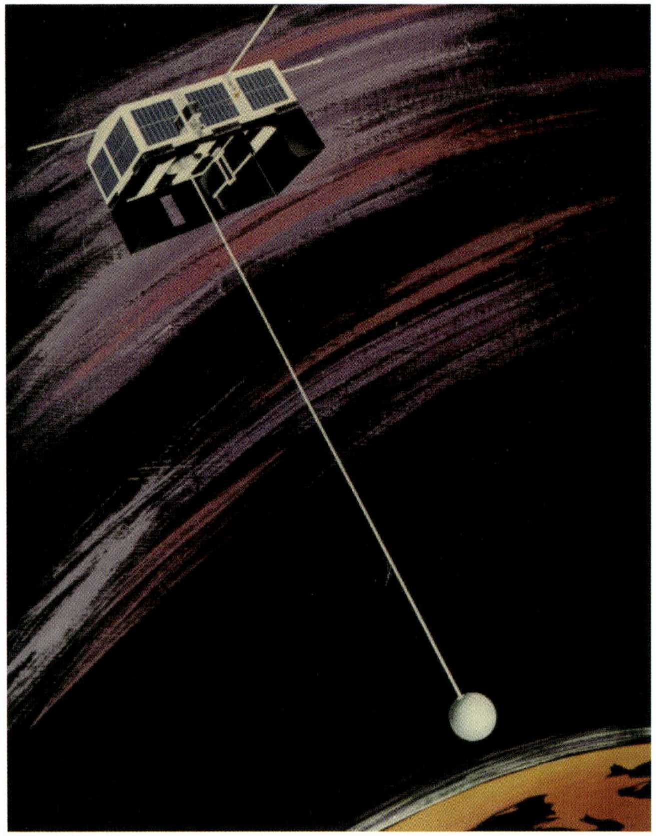

The rapidly increasing number of military and national satellites on orbit required growth in the organization supporting those satellites and conducting the network operations necessary to perform telemetry, tracking, and commanding (TT&C) functions. By 1973, the Air Force was expanding the facilities at the Air Force Satellite Test Center at Sunnyvale, California. This circa 1973 photo shows the construction of additional facilities at the U.S. Air Force Satellite Test Center. Later, the installation was renamed Sunnyvale Air Force Station. It retained that name until July 1987, when it was renamed Onizuka Air Force Base, to honor Lieutenant Colonel Ellison Onizuka, who was lost in the space shuttle *Challenger* disaster in January 1986.

The first proposal for a geostationary communications satellite was made in 1945 by Arthur C. Clarke in a letter to *Wireless World* magazine. By the mid-1950s, the U.S. Navy was involved in Operation Moon Bounce with the objective of developing a reliable form of wireless communications by bouncing radio signals off the moon. Military Satellite Communications (MILSATCOM) systems advanced quickly in the years after Clarke's proposal and by the 1980s, they played an important role in daily military operations and training. The Initial Defense Satellite Communications System (IDSCS) launched on June 16, 1966, and the DSCS II system that followed soon became the cornerstone of military satellite communications for U.S. forces. Nearly two years to the day later, on June 13, 1968, the last of the initial satellites, IDSCS 26, launched from the Eastern Space and Missile Center aboard a Titan 3C booster. Between 1971 and 1982, the Air Force launched fifteen Block II DSCS satellites.

The Defense Support Program satellites, first launched on November 6, 1970, represented an operational outgrowth from the MIDAS program that originated as a component of WS-117L. Twenty-three DSP satellites were orbited by the time of the program's last launch on November 10, 2007. The Defense Support Program has formed the backbone of the U.S. ballistic missile early warning system. DSP, then operated from Falcon Air Force Base, provided critical early warning of Iraqi scud missile launches during the 1991 Gulf War. That early warning gave forces and civilian populations time to shelter and also gave military forces the opportunity to launch counter measures. DSP satellites operate in the geosynchronous orbit belt at 22,300 miles over the equator.

CREATING A CONSOLIDATED SPACE OPERATIONS CENTER

The development and fielding of U.S. space systems, especially in support of military forces and other national functions, led rapidly to the need to expand the capabilities of the Air Force's satellite control network. As requirements increased, they stretched the capabilities of Sunnyvale Air Force Station. Meanwhile, the civilian community near the installation grew rapidly and soon encroached on the site's boundaries making future expansion problematic, to say the least. Exacerbating the encroachment issue, the Sunnyvale site also presented vulnerabilities, because it was bordered on two sides by Highways 237 and 101, respectively. Furthermore, the Air Force Satellite Control Facility, the organization operating and supporting military and national satellite systems, and its home installation, provided the Air Force's only centralized tracking, telemetry, and commanding capability. No backup capability existed! Loss of capabilities offered only at the Sunnyvale site, whether to enemy action, earthquake, or other cause, would seriously and negatively affect the U.S. ability to conduct military satellite operations.

To counter these issues, Department of Defense officials, in early 1979, approved the concept for an operations-focused installation to provide a backup control node for support of existing and planned satellite constellations (a satellite operations center (SOC)), and to house an operations support center for NASA's space shuttle from which military shuttle operations would be planned (the shuttle operations center (SOPC)). Some Air Force leaders saw the space shuttle as a potentially cost-effective method of orbiting satellites. In 1979, the DoD selected the first thirteen candidates from all services for the Military Spaceflight Engineer (space shuttle payload specialist) program to support DoD missions aboard the space shuttle. Another fourteen Air Force officers were selected in 1982, and the final five in 1985.

Approval of this concept began the history of what would become Schriever Air Force Base on the high plains of Colorado, approximately ten miles east of Peterson Air Force Base in Colorado Springs.

Over the next two years, Air Force officials evaluated many possible sites for the CSOC, eventually winnowing a list to twelve potential locations. By 1981, these twelve sites had been winnowed to two locations—Kirtland AFB in New Mexico, and Peterson AFB in Colorado Springs, Colorado. Although lacking land for expansion, Peterson AFB provided the advantages of proximity to the Space Defense Operations Center (SPADOC) at Cheyenne Mountain Air Force Station and the growing space industry along Colorado's Front Range. Kirtland AFB, however, offered surplus floor space to support administrative functions, thereby reducing some of the initial construction costs. When Kirtland was recommended to Dr. Hans Mark, Undersecretary of the Air Force, Dr. Mark believed the siting criteria did not present a clear choice of Kirtland AFB. He asked for a re-examination of the criteria and added that the CSOC should provide a SOC, the SOPC, and should also house the NAVSTAR GPS control center. This reconsideration resulted in four primary candidates, adding Luke and Malmstrom AFBs to the candidate list. However, concerns by Tactical Air Command officials on the effect the CSOC would potentially have on training eliminated Luke AFB from further consideration.

Aerospace Defense Command, located in Colorado Springs, advocated in favor of Colorado Springs citing the proximity of its planned SPADOC, which could be linked to the CSOC, and emphasized that approximately 500 acres of available land east of Peterson AFB could be acquired for $600 to $700 per acre. This parcel was outside the air traffic pattern for the Peterson AFB/Colorado Springs Airport and met the other safety requirements. Still, the second analysis favored Kirtland AFB.

Dr. Mark and the Air Force's senior military leadership raised questions about the analysis and asked General James E. Hill, Commander in Chief, Aerospace Defense Command, to study the operational considerations of the site selection. General Hill responded that the decision on location of the CSOC "went beyond the initial analyses." He pointed to the operational and organizational gains to be realized by co-locating the CSOC in Colorado Springs, reasoning that these considerations "must weigh heavily on the final decision." Ultimately, General Hill's reasoning prevailed and on December 20, 1979, Dr. Mark announced to the public that the CSOC would be located on one of three parcels of land under consideration east of Colorado Springs. Two of those parcels were owned by the State of Colorado and the other one privately held.

Ultimately, the Air Force chose parcels owned by the State of Colorado. Following negotiations, the state granted the federal government a deed to approximately 640 acres of land a few miles south of the intersection of Highway 94 and Enoch Road. On May 17, 1983, contractors broke ground on what would become Falcon Air Force Station, named for the nearby unincorporated town north of the installation. For over two years, contractors worked to complete sufficient facilities to open the base, including headquarters and operations buildings, support facilities, and infrastructure.

When the ground-breaking ceremony for Falcon AFS occurred on May 17, 1983, early estimates suggested a workforce totaling nearly 2,000 persons at full operational capability. The initial construction focused on operational and technical facilities and administrative workspaces to support those activities. Construction occurred in two phases. In the first phase, with a price tag of $6 million, the Schmidt-Tiago company, based in Colorado Springs, completed the excavation and related work for the installation, its roadways and parking lots. On March 1, 1984, three weeks after being awarded the contract for the "vertical construction" of the installation, crews from Bechtel National Incorporated of San Francisco began the second phase.

On July 8, 1985, the 2nd Space Wing activated in a ceremony at Falcon Air Force Station, although the installation was not complete enough to allow the new wing to occupy the facilities. A ribbon-cutting ceremony on September 26, 1985, symbolized the activation of Falcon AFS. Initial construction costs totaled $91,450,000, less than two-thirds of the amount appropriated. On December 1, 1985, Air Force Space Command activated the 1st Manned Spaceflight Control Squadron at NASA's Lyndon B. Johnson Space Center. This squadron would assume primary responsibility for operational planning for military space shuttle missions. Unfortunately, budget constraints and the loss of the shuttle *Challenger* in January 1986 led to the discontinuance of the Air Force's military shuttle program and the termination of the SOPC at Falcon AFS; the 1 MSCS inactivated on June 30, 1989.

Over the next two years, the 2nd Space Wing's growing team prepared the installation for the eventual turnover of system responsibilities from Sunnyvale and Vandenberg AFB (the GPS Control Center) to Falcon Air Force Station. Meanwhile, contractors continued their efforts to complete the initial facilities that would house the new satellite operations centers. Air Force Systems Command and Air Force Space Command developed a three-phase plan to transition satellite operations from the Onizuka and Vandenberg sites to the Consolidated Space Operations Center. That plan would culminate with the final turnover of the communications segment in the summer of 1993, symbolized by a final turnover ceremony on September 27, 1993.

By 1987, the host 2nd Space Wing's satellite control crews already had begun limited operations for some satellite systems. Much of the daily operations were related to training crews on the systems and conducting operations under the supervision of contractor or Air Force Systems Command personnel. Operational turnovers of specific satellite systems were based on the wing's ability to meet established criteria for crew training and readiness, hardware and software capability and support, and other factors agreed upon by the working groups managing the processes.

The establishment of a military installation brought with it opportunities for commercial and residential development and economic growth. Several private companies including, by 1986, Nova Tech (an engineering company and defense contractor) and Forest City Enterprises (a real estate development company) eyed the area for development. At that time, Nova owned several large parcels near Falcon AFS and sought to purchase others for the development of the Nova Technological Center. Aerospace Centre and East Glen were other local developers proximate to the installation. Meanwhile, the Colorado State Board of Land Commissioners negotiated with Forest City Enterprises for the development of 16,000 acres of state land surrounding the new installation. The state's master plan called for a mixed development of residential, commercial, industrial, and research enterprises.

This proposed development surrounding Falcon concerned Air Force officials. While local and headquarters commanders expressed support for some narrow development around the installation, they believed it should be limited to small ranchettes and agricultural uses such as grazing. These activities would be compatible with the level of security base leaders desired and prevent encroachment or interference with existing and planned mission expansions—issues that had hindered expansion at Onizuka Air Force Station. Air Force officials also sought to obtain additional land to create sufficient easements and buffers to prevent future mission-hampering encroachment. By February 1986, negotiations on land transfers began with the State of Colorado. These negotiations, combined with purchases and leases of privately owned parcels surrounding the installation, resulted in the Air Force's acquisition of over 2,000 additional acres. This provided the base with room to expand while maintaining an adequate buffer against encroachment.

The new installation also provided an opportunity for an expansion of the Air Force Satellite Control Network under the second phase of the automated remote tracking station (ARTS) upgrade. That contract provided for construction of the Colorado Tracking Station, call sign PIKE, at the new base. PIKE's location in the interior was unique compared to other remote tracking stations, most of which were on islands or near the coast. It's collocation with the second AFSCN control node allowed the site to be hard-wired to the network, which provided additional communications security.

Cost-cutting measures and international factors in 1995-1996 led to the closure of the AFSCN's Indian Ocean Tracking Station. Ten years later, budget issues led to a reduction at the Colorado Tracking Station from 24/7 operations to 40 hours per week—a first for a remote tracking station. Then, in 2012, additional budget cuts within the Department of Defense necessitated the closure of PIKE. Crews conducted PIKE's final support on July 9, 2012. Two years later, in September 2014, following the removal and transfer of PIKE's antenna and some related equipment, crews began demolition of the site.

When Desert Shield began following the Iraqi invasion of Kuwait, Falcon AFB, as it was then named, was in this period of evolution. Many of the systems operated at the installation today by the 50th Space Wing were transitioning from other operational sites. The 1st Satellite Control Squadron (now named 1st Space Operations Squadron) had assumed operational responsibility for the Defense Support Program satellites that provided theater missile warning. The 2nd Satellite Control Squadron (now 2 SOPS) operated some Global Positioning System satellites, and the 3rd Satellite Control Squadron (now 3 SOPS), which activated on February 2, 1990, operated some military satellite communications systems, including the Defense Satellite Communications System. The 2nd Space Wing's 1000th Space Operations Group operated the Defense Meteorological Satellite Program, and the 2nd Satellite Tracking Group, operated the Air Force Satellite Control Network from Onizuka AFB supporting these and other national satellite systems. Each of these systems played an important part in Desert Shield/Desert Storm.

GPS and space-based navigation had been available to some U.S. forces for nearly a decade before 1990. However, this availability was limited and consisted chiefly of the Block I experimental GPS satellites. The operational satellites, Block II, did not begin launching until 1989. By the time of the Gulf War, a variety of Air Force aircraft and some mostly experimental munitions were GPS equipped but Army and Marine Corps ground units were not well equipped. The Army, during its buildup, acquired nearly 5,000 GPS receivers, and troops were augmenting those with their own personal purchases. When the Gulf War started, the GPS constellation was only two thirds complete. Nevertheless, it provided invaluable positioning and navigation information to U.S. and allied forces.

In the run-up to the initiation of combat operations, the Air Force launched two GPS satellites on October 1 and November 26, 1990, and the 1st and 2nd SCS crews completed their actions to make them operationally available prior to the initiation of combat operations. At the 2nd Space Wing, the 2 SCS maintained nearly 100 percent operational effectiveness. Crews also returned a non-operational satellite to service to provide additional capability. Soldiers and Marines bought their own GPS receivers and found ways to mount them on their vehicles and tanks. GPS made bombing and artillery fire much more accurate and reduced fratricide (friendly fire) incidents. Additionally, GPS contributed to the land forces' ability to move quickly and at night with precision and for those forces to be supported with food and supplies.

The ground war was over in 100 hours! Space-based capabilities—positioning and navigation, space-based weather forecasting, and satellite based communications—each contributed to the success of the air campaign that crippled Iraqi infrastructure and contributed to the rapid defeat of Iraqi ground forces.

The Defense Support Program provided U.S. and coalition forces with advanced warning of Iraqi Scud attacks and provided opportunity for countermeasures and personnel sheltering. The DMSP weather capability, also operated by the 2nd Space Wing, played an important part in developing and implementing operational plans and the daily air combat plan. Targets, aircraft, and weapon loads could be selected based on highly accurate forecasts from space-based platforms. Military satellite communications (MILSATCOM) capabilities provided the by 3rd SCS were an integral part of campaign planning and execution.

Desert Storm has been characterized by some military experts as the first space war. Indeed, it was the first time all U.S. military satellite systems were used extensively to support air, land, and sea operations during a conflict. The theater missile warning provided by DSP which was operated by the 1st SCS limited the effectiveness of Iraq's scud missile attacks and saved many lives, coalition, U.S. Forces, and civilian.

Lessons learned during the conflict affected Schriever AFB in ways seen and unseen. Some of those lessons occasioned the assignment of new organizations and programs to the base, while others principally affected the way in which the wing and its components conducted their missions. Changes to procedures and practices and the internal realignment of some unit functions sought to improve the wing's and the base's response to contingencies and its delivery of decisive global effects.

By the 1970s, barely twenty-five years after the United States launched its first satellite, the only Air Force facility for satellite tracking, telemetry, and commanding, the Air Force Satellite Test Center, near Sunnyvale, California, had nearly reached its operational capacity. Encroachment from the civilian community presented security concerns and prevented the physical expansion necessary for the installation to support new systems that were coming online. By the end of the decade, the newest satellite system, Global Positioning System Block I experimental satellites, were being launched and controlled from mobile facilities at Vandenberg Air Force Base.

The Air Force selected a site east of Colorado Springs as the home of the new Consolidated Space Operations Center. The site, situated about ten miles east of Peterson AFB, south of the intersection of state Highway 94 along Enoch Road in El Paso County, consisted of 640 acres of land obtained from the state in exchange for federal land holdings near Denver. The location had not been considered in the initial siting studies, which prompted allegations that the choice had been political. The location, however, provided the Air Force with opportunities to counter the kind of encroachment that had limited expansion of mission activity at the Satellite Test Center, near Sunnyvale, California.

On May 17, 1983, contractors and Air Force officials ceremonially broke ground on the new Consolidated Space Operations Center and Falcon Air Force Station. Built under the supervision of the Army Corps of Engineers, the Air Force Systems Command was originally the agency responsible for construction of the installation, while Air Force Space Command would serve as the command responsible for operational tasks. This partnership remained until the completion of mission transitions and final turnover of the CSOC to Air Force Space Command in September 1993.

In this fall 1983 aerial view, a nearly completed Falcon Parkway appears in the foreground. The image also depicts construction offices, graded areas for the continuation of Falcon Parkway, the primary parking lot, and graded areas for the primary operational facilities in the upper left corner.

Construction of the physical installation progressed quickly, as shown here in this undated image. Looking to the west, in the background (right), is building 300, which eventually would serve as the installation headquarters facility and administrative offices for some of the satellite control (later renamed space operations) squadrons. Left of building 300 is building 400, destined to become the primary operations building and home to the satellite operations centers from which the base's primary missions would be executed. The right center facility is building 500, which initially served as the base headquarters and administrative offices. In the center left, the base's central power plant is visible.

On September 26, 1985, a ribbon cutting ceremony marked the opening of Falcon Air Force Station. The 2nd Space Wing, which was activated in July and had been operating from offices at Peterson AFB, began moving into the new facilities. Construction of administrative, operational, and technical work areas, however, continued, and the 2nd Space Wing's cadre of about 200 personnel primarily worked in building 500.

The opening of Falcon Air Force Station was the first step in the development of the Consolidated Space Operations Center. The construction of necessary infrastructure, including operational centers and other facilities continued well into the late 1980s. By May 1987, most of the initial operations-focused facilities were complete, as was the Colorado Tracking Station, call sign PIKE. The Colorado Tracking Station had been contracted during the second phase of the Automated Remote Tracking Station (ARTS) program upgrade to the Air Force Satellite Control Network. Construction of the PIKE facility began in 1984, and the station became operational in 1989.

Beginning in 1987, the Air Force Systems Command, following detailed and continuously updated plans developed with Air Force Space Command and the 2nd Space Wing, began the long process of transitioning mission operations for satellite tracking, telemetry, and commanding, as well as Air Force Satellite Control Network operations, from California locations to Falcon AFS. Global Positioning System operations were among the first to begin transitioning, with crews of the 1st and 2nd Satellite Control (later Space Operations) Squadrons conducting aspects of the GPS mission in 1987. The 2nd Satellite Control Squadron activated on Oct. 1, 1985, and personnel began relocating from the GPS operations center at Vandenberg AFB. The 1st Satellite Control Squadron activated on Oct. 5, 1987, to assume launch, early orbit, and anomaly resolution responsibilities. Operational turnover of the first GPS vehicle occurred on Dec. 21, 1988. Formal turnover of the GPS system to Air Force Space Command and the 2nd Satellite Control Squadron occurred on April 24, 1990. The base's building 400, named later to honor astronaut and Colorado Congressman-elect John "Jack" Swigert, housed the satellite operations centers.

As the new wing continued to expand knowledge and experience within its satellite crew force, the role of the 1st Satellite Control Squadron (later redesignated 1st Space Operations Squadron) changed as personnel trained in specific systems began to populate the 2nd and 3rd Satellite Control Squadrons. On February 23, 1989, the 1st SCS conducted its first transfer of satellite control authority for a Global Positioning System satellite to the 2nd Satellite Control Squadron. Over the next weeks, as the 1st SCS gained authority for the remaining nine GPS satellites, crews completed their check and began the process to transfer control to the 2nd SCS. The 1st retained some responsibility for GPS operations, including launch support, anomaly resolution, and disposal activities. It would continue these operations until the installation and operational acceptance of the Launch, Anomaly, and Disposal Operations (LADO) upgrade to the 2nd Space Operations Squadron's command and control architecture.

On February 23, 1989, the 1 SCS transferred satellite control authority for GPS satellite II-1 to the 2 SCS.

GPS II-1 launched on February 14, 1989 and was the first GPS launch controlled from Falcon AFB and the first launch of the Block II operational satellites.

Defense Satellite Communications System Block II

Defense Support Program

Defense Meteorological Satellite Program

The 1st Satellite Control Squadron also assumed some operational responsibilities for the Defense Support Program, Defense Meteorological Satellite Program, and even initially the Defense Satellite Communications System. The 1st SCS then transitioned control to other squadrons, eventually winnowing its responsibilities to DSP operations and launch, early orbit, and anomaly resolution support for GPS operations. The 3rd Satellite Control Squadron, activated on Feb. 2, 1990, quickly assumed operational control of the DSCS system, adding to their mission command and control of the U.S. Navy's Fleet Satellite (FLTSAT) and the NATO III communications programs. Later, the unit would add the United Kingdom's Skynet communications satellite to its list of supported programs.

Falcon
FEEDBACK
Volume 3, Number 16 | Falcon AFB, Colo. | Feb. 7, 1992

50th SPW stands up

The 50th Space Wing activated Jan. 30 in a ceremony attended by more than 500 people in the Satellite Dish dining facility.

Lt. Gen. Thomas S. Moorman Jr., commander of Air Force Space Command, presided over the ceremony, which also saw the inactivation of the 2nd Space Wing and the activation, inactivation or redesignation of both wings' various subordinate units.

"Today, we of the old 2nd Space Wing proudly take on the more than 50 years of history, lineage and traditions of the 50th Tactical Fighter Wing," DeKok said. "We build on a proud past as we, the new 50th Space Wing, take the 50th Tactical Fighter Wing's motto, 'Master of the Sky,' to even greater heights.

Prior to inactivating the 2nd SPW, Moorman also presented DeKok and the wing the Air Force Outstanding Unit Award, congratulating the wing's people for their work supporting the country's space program and the extra satellite operations requirements and challenges of Desert Shield.

Photo by Sgt. George Hayward

Bldg. 300 gets with the times, trading its "2nd" sign for the "50th" moniker, courtesy of Amn. Timothy Russom of the 50th Civil Engineering Squadron.

Photo by Sgt. George Hayward

50th Space Wing Commander Col. Roger DeKok unfurls the 50th Tactical Fighter Wing's highly-decorated flag.

INSIDE

Ultimately, the 2nd Space Wing, Falcon AFB's host unit, managed assets around the globe. Its 2nd Satellite Tracking Group headquartered at Onizuka AFS, California, the former home of the Air Force Satellite Control Facility, operated and managed the various world-wide remote tracking stations. From Offutt AFB, Nebraska, the 1000th Space Group operated stations at Washington's Fairchild AFB, and Maine's Loring AFB, supporting the Defense Meteorological Satellite Program. A squadron at Onizuka, the 5th Satellite Control Squadron, also supported the DMSP mission. At Falcon AFB, the wing's three satellite operations squadrons operated the Global Positioning System, the Defense Support Program, and a host of wideband satellite communication systems.

The 2nd Space Wing, however, would not see its labors come to fruition. Prior to the formal turnover of the Consolidated Space Operations Center on Sept. 27, 1993, the 2nd Space Wing inactivated on January 30, 1992, and was replaced on that day by the 50th Space Wing, redesignated from 50th Tactical Fighter Wing. The 50th, "Master of the Sky," had served in Europe for the preceding thirty-eight years, earning a distinguished history. To maintain that history, the 50th was chosen to activate at Falcon AFB and the "Master of the Sky" became the "Master of Space."

Indicative of the growing significance of the small base on the Colorado plains, Air Force Space Command activated, effective April 30, 1992, the 4th Strategic Reconnaissance Squadron, Medium, having previously redesignated it 4th Space Operations Squadron. The 4th would prepare for and then operate the Milstar communications satellite system, exercising platform control and communications payload management. Manufactured by Lockheed, Milstar was a joint service system that provided secure, jam-resistant, worldwide communications to meet the essential wartime requirements for high-priority military users. Each of the planned Milstar satellites served as a "smart" switchboard in space, directing traffic from terminal to terminal anywhere on Earth. The 4 SOPS' Milstar mission included a mobile capability to ensure continued service. Nearly two decades later, the squadron also would gain operational responsibility for the Advanced Extremely High Frequency (AEHF) satellite system.

In 1992, the 1st Space Operations Squadron was leading the 50th Space Wing's efforts to convert many space operations billets from officer to enlisted. This multi-year program eventually encompassed all of the wing's space operations squadrons and was projected to save approximately $10.5 million (in fiscal year 1991 dollars) in manpower costs as the wing converted nearly 200 officer billets to enlisted space operators. The squadron made news again in October 1994, when Technical Sergeant Kathy Rogers (center) was appointed as the wing's first enlisted space operations flight chief. On December 22, 1994, an all enlisted crew consisting of Technical Sergeant Reginald Logan, Staff Sergeant Marvin Gielen, Jr., and Senior Airman Clay Hyink, completed first acquisition for a Defense Support Program satellite launch.

Eight years after Falcon Air Force Station (later Falcon Air Force Base, Schriever Air Force Base) opened, the installation and its workforce were prepared to accept the final operational transfer to complete the transition of assigned missions to the Consolidated Space Operations Center. During a September 27, 1993, ceremony, representatives from the Air Force Materiel Command, Air Force Space Command, and other organizations witnessed the formal transfer of the communications segment of the Air Force Satellite Control Network to the 50th Space Wing.

FALCON FEEDBACK

Volume 5, Number 12 | Falcon AFB, Colo. | Oct. 1, 1993

Turnover marks milestone for AFSPACECOM, 50th SPW

by 2nd Lt. Almarah Uwzayaz
50th Space Wing Public Affairs

The Consolidated Space Operations Center at Falcon AFB was officially turned over to Air Force Space Command Monday.

AFSPACECOM now assumes control of CSOC, along with the formal responsibility and accountability for its development and operation, from Air Force Materiel Command — the command that has controlled this innovative idea since conception.

"This ceremony marks the culmination of a long journey for Air Force Materiel and Space Command," was relayed in Brig. Gen. Eugene Tattini's, Headquarters Space and Missile Systems Center, remarks.

In 1979, the Secretary of the Air Force directed the development of a second satellite control node for Defense Department satellite operations. At that time, DOD satellites were being controlled from the Air Force Satellite Control Facility at Sunnyvale, Calif. This is when Brig. Gen. Roger DeKok, AFSPACECOM Plans and Programs director, began his association with what has become Falcon AFB.

"It is unusual to play a part in the planning stages of a program, help guide it through 14 years of planning, developing and testing, have the privilege to command its various components and then accept its turnover," DeKok said. "Materiel Command was the developer and Space Command its operator — but the people here on Falcon have made it happen."

Today, CSOC is the primary control node for all Air Force satellites.

But, this turnover isn't the first in the CSOC program.

The first turnover milestone was reached in 1986 when the facilities' segment was given to the 50th Space Wing. Since then, the Weather Support Unit was turned over in 1989 and the Security Control System in 1991. Monday's ceremony was for the Communications Segment, Satellite Operations Complex and Network Control Segment, and signified the completion of component turnover.

Photo by TSgt. Pat Powers

(From left) Col. Gregory Gilles, 50th Space Wing commander; Lt. Col. Nasaira LeBlanc, Headquarters Space and Missile Systems Center Consolidated Space Operations Center program manager; and Brig. Gen. Roger DeKok, Headquarters Air Force Space Command director of plans, reaffirm Air Force Materiel Command's commitment to deliver the CSOC to AFSPACECOM and the 50th SPW by signing the transfer document

CSOC segments perform satellite tracking, telemetry data analysis, and command and control services for orbiting satellites in the assigned programs. Planning, resource readiness testing, launch and early orbit, communications connectivity, and anomaly resolution functions for assigned Department of Defense satellites are also accomplished.

Based on the demonstrated test performance conducted by the Air Force Operational Test and Evaluation Center from mid-May through mid-August, the CSOC was determined to be operationally effective and suitable.

The CSOC system also received a high contact success rate that stemmed directly from the crews' abilities, training and the use of established procedures.

"This ceremony [and it's subsequent turnover] should be a source of pride for all of you involved in this dynamic and successful CSOC program," said Col. Gregory Gilles, 50th SPW commander.

On November 1, 1994, Air Force Space Command accepted from Air Force Materiel Command satellite control authority for the first of the Air Force's newest satellite systems, Milstar Flight 1. Air Force Space Command then delegated that authority to the 50th Space Wing and the 4th Space Operations Squadron. Milstar Flight 1 launched on February 7, 1994.

Milstar provided the President, Secretary of Defense, and the U.S. Armed Forces with assured, survivable satellite communications, with a low probability of intercept and detection. The system was designed to overcome enemy jamming and nuclear effects and was DoD's most robust and reliable satellite communications system.

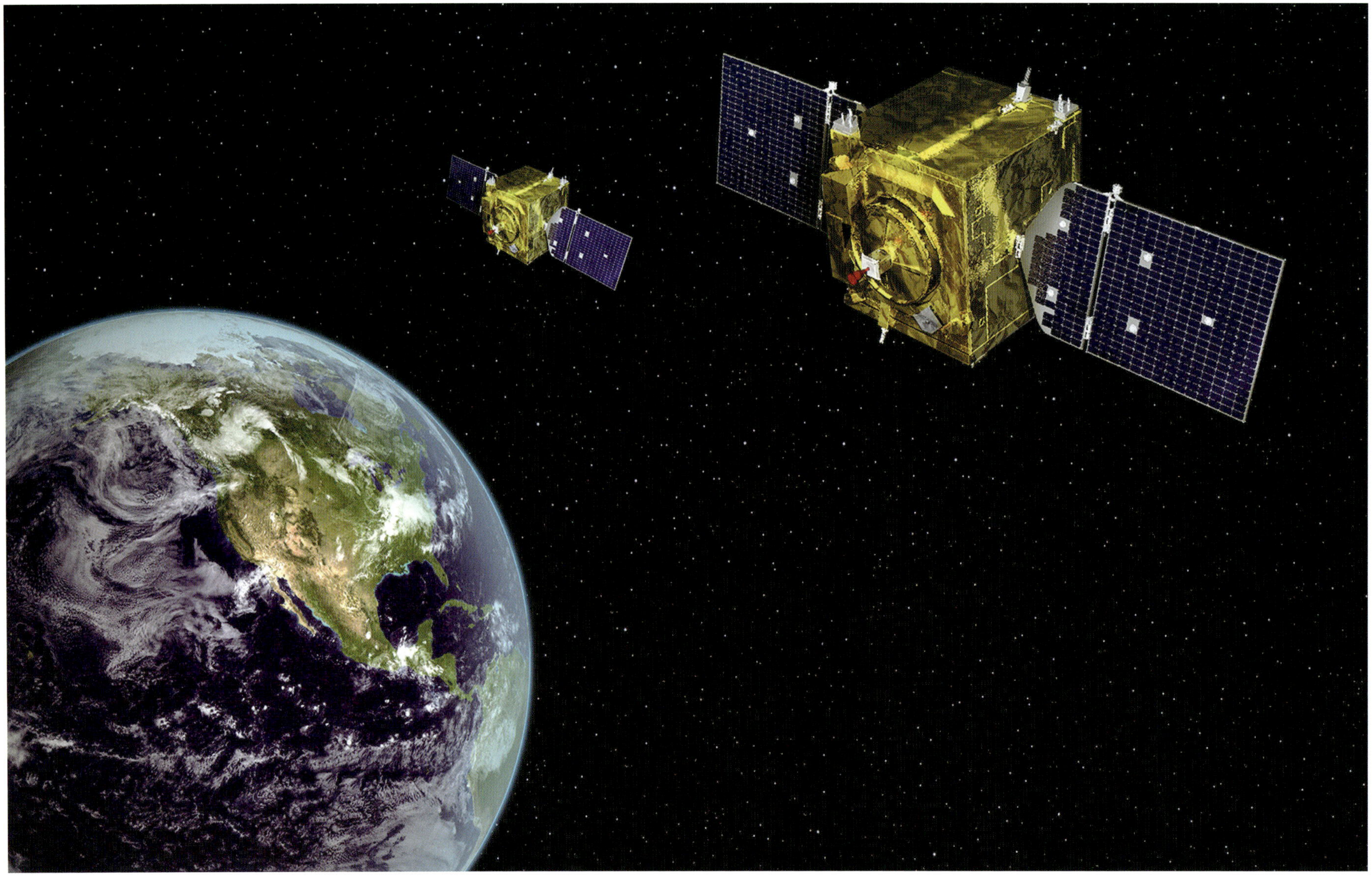

General John E. Hyten, Air Force Space Command commander, declared the Geosynchronous Space Situational Awareness (GSSAP) satellites operational on September 29, 2015 following nearly a year of on-orbit testing. The 50th Space Wing's 1st Space Operations Squadron began GSSAP operations following this announcement, continuing mission growth at Schriever AFB.

NEW MISSIONS, NEW UNITS: FALCON AFB GROWS

Activation of the 50th Space Wing (formerly 50th Tactical Fighter Wing) to operate Falcon AFB and assume its primary mission responsibilities, while important in the base's history, was less significant than other actions that preceded and followed it. In the waning years of Ronald Reagan's second term as president and during the George H. W. Bush Presidency, national space policy included emphasis on the possibilities of continental defense from space-based platforms. Early in the construction of Falcon AFS, the United States determined to locate the Joint National Test Facility, the research laboratory for President Ronald Reagan's Strategic Defense Initiative (SDI) at the new installation. Construction of the JNTF (later renamed Joint National Integration Center (JNIC)) began in 1987 and was completed in 1990, in time for a visit by British Prime Minister Margaret Thatcher. The installation's location, proximate to the growing "space corridor" along Colorado's Front Range of the Rocky Mountains while still secluded from the growth of the Colorado Springs metropolitan area, provided the same protection against encroachment that had been a factor in the site's selection as the home of the CSOC. The availability of additional land parcels at low cost, as well as the relatively inexpensive cost of utilities also were factors. Additionally, the site's relative remoteness afforded a level of operational security lacking at other potential locations.

By the end of George H. W. Bush's administration, interest in and funding for SDI declined. The Missile Defense Agency's (MDA's) Missile Defense Integration and Operations Center (MDIOC) joined the Falcon Air Force Station team in 1988 as the SDI Organization (SDIO) and shared facilities in the Jack Swigert Space Operations Facility until completion of the Joint National Test Facility in 1990. Secretary of Defense Les Aspin renamed the organization as the Ballistic Missile Defense Organization in 1993, and a major reorganization followed. The focus of the organization also changed over the course of time, from one centered on defense against a Soviet ballistic missile attack to concerns about a missile attack by rogue states or missile-capable terrorist organizations. In 1999, the JNTF was renamed JNIC and later renamed Missile Defense Integration and Operations Center.

The agency's joint service team supported research and development, system-level tests and evaluation, and provided operational and training support to U.S. combatant commands. The Integration and Analysis Operations Center assisted in the integration of sensors and weapon systems by developing and implementing missile defense command, control, battle management, and communications capabilities. The center developed models and simulations for use in testing and training and provided support to exercises, system engineering, and analysis.

Through the 1980s and into the 1990s, Falcon AFS expanded. Continuing developments in space-related technologies added importance to Falcon's consolidated space operations and brought new organizations and missions. Reflecting the installation's growth in function and form, Falcon AFS was renamed Falcon Air Force Base on June 13, 1988. While the installation's physical character expanded, so did its mission activity.

Throughout the 1990s, the installation's organizations continued to develop and mature. As the Global Positioning System moved closer to initial operational capability, the Air Force activated Detachment 46 of the Air Force Technical Applications Center at Falcon AFS to support the Nuclear Detonation Detection System payload carried onboard GPS satellites. "Det 46", as it was informally known, opened its operations center on November 8, 1991. Planning for new missions also added units to the 50th Space Wing. Air Force Space Command activated the 4th Strategic Reconnaissance Squadron, effective April 30, 1992, and redesignated it the 4th Space Operations Squadron. The squadron would conduct the Milstar protected satellite communications mission when that system was fielded, beginning in 1994. Later, the squadron also assumed operational responsibility for the Advanced Extremely High Frequency (AEHF) satellite program that began launching in 2010 and would serve ultimately as a follow-on system to Milstar. By 2015, 4 SOPS crews operated five Milstar and three AEHF satellites and their communications payloads. They were responsible for "cradle to grave" operations and management of these critical communications programs supporting

U.S. and allied forces in the field and in garrison, and U.S. national command authorities.

Implementing lessons from Desert Shield/Desert Storm, the units of the 50th Space Wing were assigned mobilization missions, requiring those squadrons to deploy, when called, teams or individuals to augment forward forces. This new task required the wing to complete a variety of tasks, including specialized training for deployed forces, building and maintaining mobility bags in several different configurations for specific contingencies and environmental conditions, and developing processes and procedures for moving personnel "downrange." Until 2003, the 50th relied on assistance from Peterson AFB to meet these new tasks. However, the June 1, 2003, activation of the 50th Supply Squadron as the 50th Logistics Readiness Flight provided the wing and the installation with an inherent logistic function to manage these requirements.

Air Force Space Command activated the Space Warfare Center (later redesignated Space Innovation and Development Center (SIDC)) on November 1, 1993, and charged that organization with incorporating space capability in major command, theater, and campaign war plans ... simulation, modeling and war gaming activities and to lead development efforts to improve combat applications of space systems. Even before the center's formal activation, General Anthony McPeak, Air Force Chief of Staff, directed that the Space Warfare Center would oversee the Air Force Tactical Exploitation of National Capabilities (TENCAP).

For nearly two decades, SIDC conducted operations at Schriever AFB, expanding its responsibilities and organizational construct to meet new requirements and missions. The SIDC added the Space Battlelab (1997-2007), the 17th and 14th Test Squadrons (under the 595th Space Group), the Space Operations School (2000-2004) and the Advanced Space Operations School (2009-2013). The Center's 595th Space Group also included the 26th and 527th Space Aggressor Squadrons, which provided adversarial forces for battle simulations. The 3rd Space Experimentation Squadron (3 SES) joined the SIDC in 2006 and in 2013 transferred to the 50th Operations Group as part of the reorganization that included the inactivation of the Space Innovation and Development Center and reassignment of its remaining component units to Air Force Space Command or to the United States Air Force Warfare Center.

In January 2012, SIDC leadership began the task of reducing the size of the organization. Resource Management Decision (RMD) 703 mandated a reduction of the Air Force civilian workforce to 2010 levels. Over the next fifteen months, the SIDC transferred most of its missions and organizations to the USAF Warfare Center (USAFWC). The SIDC inactivated on April 1, 2013. General William Shelton and Colonel Kevin Rhoades oversaw furling of the unit flag.

On October 1, 1994, Air Force Space Command activated the 11th Space Warning Squadron at Falcon AFB based on lessons learned from Operation Desert Storm. Existing missile early warning systems were accurate, but were not intended for theater use. Commanders identified the need for timely and accurate details on short-range ballistic missiles, such as the Scud variant used by Iraqi forces. In response, the Air Force initiated Talon Shield, under the TENCAP organization to find better methods of processing and disseminating missile warning data to forces. The result of this effort was the Attack and Launch Early Reporting to Theater (ALERT) system. The squadron operated as a component of the 21st Space Wing, headquartered at Peterson AFB, until it inactivated on December 31, 2002. On December 3, 2007, the unit again activated at Schriever AFB, this time as a component of the 460th Space Wing, which was headquartered at Buckley AFB, Colorado.

In 1995, and again in 2005, the Base Realignment and Closure Commission (BRAC) recommended and the U.S. Congress, via legislation signed by the president, implemented actions that changed the landscape of Schriever AFB. Onizuka AFB's encroachment by commercial and private development led in 1995 to recommendations to shift some operational activities from there to Falcon AFB. Over the next several years, Onizuka-based units inactivated and Falcon units assumed greater responsibility for AFSCN operations. The 22nd Space Operations Squadron assumed responsibility for oversight of the network's remote tracking stations in 1997, following inactivation of the 750th Space Group at Onizuka AFB. That transition made the 50th Operations Group responsible for nearly all aspects of the satellite TT&C functions conducted by the 50th Space Wing. By 2010, Onizuka AFB's remaining operations transferred to other locations, as a result of the 2005 BRAC process. The base's remaining unit, the 21st Space Operations Squadron, relocated its operations to the newly completed Ellison Onizuka Satellite Control Facility at Vandenberg AFB, California.

Meanwhile, the 50th Space Wing's communications organizations underwent a series of organizational realignments in the late 1990s and early 2000s as the wing sought to refine operational effectiveness and efficiency while ensuring that communications, a critical aspect of wing operations, remained under the management of a colonel. By March 2004, Falcon's original communications unit, the 1879th Communications Squadron, had been activated and redesignated 50th Network Operations Group. Assigned to the group were the 50th Space Communications Squadron and the 21st, 22nd, and 23rd Space Operations Squadrons, which transferred from their previous assignment with the 50th Operations Group. The "NOG", as it came to be known, assumed responsibility for all AFSCN operations and maintenance and, later, also assumed responsibility for supporting GPS ground antennas. On October 1, 2003, Air Force Space Command activated the 50th Comptroller Flight, redesignating it 50th Comptroller Squadron on July 1, 2004.

Also in 1995, the United States Naval Observatory (USNO), working with the 50th Space Wing and other Air Force officials, began efforts to relocate the Alternate Master Clock (AMC) from Richmond, Florida, to Falcon AFB, where its collocation with the GPS Master Control Station, the primary means of global time transfer, provided several benefits, including more robust backup capability and a more secure operating environment. At Falcon, the clock achieved initial operational capability in November 1995 and became fully operational on July 23, 1996. From that point, engineers sought to incorporate the precise time the clock provided to the AFSCN, GPS, and some TALON programs under the purview of the Space Warfare Center (later Space Innovation and Development Center). The ability to provide precise timing to combatant forces enhanced the capabilities of those forces in navigation, weapons delivery, communications, and intelligence.

Planning for a new AMC at Schriever AFB, however, began within a few years. The nearly ten year development and planning process culminated with the arrival of a new clock from the USNO on December 4, 2012. Moving the AMC to its new home in the GPS Master Control Station from the loading dock required that it be floated across a series of mats to minimize potential damage to the clock's components. Colonel James P. Ross, remarked that the new clock represented "a quantum leap in the accuracy of the . . . Alternate Master Clock, which continually provides the world with precise timing via the Global Positioning System."

Preceding the USNO at Schriever was the Naval Satellite Operations Center (NAVSOC), Detachment Delta. Detachment Delta, originally activated to serve as a liaison between the 3rd Space Operations Squadron and NAVSOC, provided the NAVSOC with backup capability supporting the primary operations center at Point Mugu, California, which operated a variety of naval communications satellite systems, including the Ultra-High Frequency Follow-On (UFO), Fleet Satellite Communications System (FLTSAT), and the Mobile User Objective System (MUOS).

Detachment 46, Air Force Technical Applications Center joined the Falcon AFB team on September 24, 1996. Among its primary activities, the detachment was charged with monitoring the Nuclear Detection System (NDS) sensor carried aboard GPS satellites, examining its state-of-health information, detecting anomalies, and recommending corrective actions, when observed, to the 2nd Space Operations Squadron.

On September 4, 1997, the 310th Space Group (later 310th Space Wing) joined Schriever AFB's growing list of organizations and became the first space group in the Air Force Reserve Command. However, Air Force Reserve Command support to space operations at Schriever began four years earlier with the activation of the 7th Space Operations Group as a component of the 302nd Airlift Wing at Peterson AFB. The 310th Space Group's first permanent facility on Schriever was completed in 2004. The 310th originally included the 7th and 19th Space Operations Squadrons, whose crews worked closely with crews of the 50th Space Wing operating assigned satellite systems. For the 19th SOPS, the assigned system was the Global Positioning System. The 7th Space Operations Squadron, meanwhile, served as the reserve associate of the 50th Space Wing's 1st Space Operations Squadron, operating at first the Defense Support Program and, then, pathfinder systems before programmatic changes resulted in the transition of the 1st and 7th SOPS to operate the Space-Based Space Surveillance (SBSS) system that provided satellite-based space situational awareness. The 310th has grown since that time and was redesignated the 310th Space Wing in March 2008. The 310th Space Wing's 6th Space Operations Squadron, formerly a component of the 50th Space Wing, provides Air Force Reserve support to the Defense Meteorological Satellite Program, while the 8th Space Warning Squadron operated, with the 2nd Space Warning Squadron, the Space Based Infrared System (SBIRS) for missile warning and defense from Buckley AFB near Denver, Colorado, while Detachment 1, 8th Space Warning Squadron partnered with Schriever AFB's 11th Space Warning Squadron, to operate the SBIRS Highly Elliptical Orbit (HEO) constellation.

The wing's mission was to deliver optimized, scalable combat-ready forces capable of delivering space and cyberspace power for faster, more lethal, more accurate effects, every day, everywhere. In fulfilling that mission, the 310th also included squadrons that were associated with the active components of the former 595th Space Group, including the previously mentioned 14th Test Squadron, an associate of the active-duty 17th Test Squadron. The 379th Space Range Squadron similarly served as a reserve associate to the active-duty 25th Space Range Squadron.

Indicative of the growth of the 310th Space Wing in form and function, construction of a new headquarters facility began. Later named in honor of Colonel William M. Bower, contractors completed the facility in 2011.

The Joint Forces Component Command (JFCC) for Integrated Missile Defense (JFCC-IMD) began operations at Schriever AFB in January 2005. A unit of the United States Strategic Command (USSTRATCOM), the JFCC-IMD was represented in the Missile Defense Integration and Operations Center and its headquarters was located there. The Joint Functional Component Command for Integrated Missile Defense supported U.S. Strategic Command's Unified Command Plan missions for synchronizing operational-level planning and global missile defense operations. The unit was accountable for supporting the development of global effects for the Department of Defense and, when directed, provided alternate missile defense execution support. It also integrated and coordinated U.S. missile defense systems and operations, synchronized theater missile defense plans into the overall global missile defense campaign, and conducted day-to-day operations and coordinated activities with associated combatant commands, other USSTRATCOM JFCCs, and the efforts of the Missile Defense Agency.

Meanwhile, Colonel (later General) John E. Hyten, 50th Space Wing commander, developed a concept for a Wing Integrated Operations Center. This center, staffed by representatives of each of the wing's primary mission and key support organizations would serve as an information collection, analysis, and dissemination center allowing improved communications flow across the wing. Equipped with this information, units at Schriever and around the globe could initiate coordinated responses to events, from adversarial attempts to interfere with wing operations, to terrorism threats, to natural disasters and weather events potentially affecting wing mission activity. His successors at the helm of the wing carried his vision forward, although the reality of funding and facility constraints eventually resulted in a significant de-scoping of the project. By 2010, the concept had been reduced to an integrated operations floor, termed the Integrated Operations Environment (IOE), for MILSATCOM operations.

When opened for MILSATCOM operations, following the occupancy of the facility by the wing's

3rd and 4th Space Operations Squadrons, the IOE quickly validated Colonel Hyten's vision. Collocated in the IOE, the two squadrons shared critical information and took advantage of each other's expertise to improve operational effectiveness. The transition to operations within the IOE was smoothed by several months of collocated operations in one satellite operations center while the build-out of the IOE occurred. During this period, the two squadrons were able to identify issues and challenges and to develop processes, procedures, and understandings to resolve those challenges. The IOE quickly became a featured operations center during visits by civic, Air Force, and congressional leaders.

While wing leaders worked to refine and implement Colonel Hyten's vision, other changes in mission tasks meant new opportunities for wing units. By the early 2000s, the 1st Space Operations Squadron was operating several experimental systems, including the Midcourse Space Experiment (MSX) and Advanced Composition Explorer (ACE). Previous to this, officials had proposed instituting a multi-mission satellite operations center (MMSOC) that would allow satellite TT&C from a common ground system architecture. The 50th Space Wing's MMSOC opened on October 29, 2006. Operationalizing the center, however, required that 1 SOPS, which would operate the various systems assigned to the MMSOC, divest its DSP operations and support to GPS launch and early orbit activities. Squadron personnel transferred DSP operational responsibility to the 2nd Space Warning Squadron at Buckley AFB, Colorado, on August 29, 2006, while the 2nd Space Operations Squadron, equipped with the new Launch, Anomaly, and Disposal Operations (LADO) software suite, assumed "cradle-to-grave" operational responsibility for GPS. In later years, 1 SOPS gained operational responsibility for the Space-Based Space Surveillance (SBSS) mission (2010) and would also gain the Geosynchronous Space Situational Awareness Program satellites when they became operational. In 2013, the squadron turned over the MMSOC and those pathfinder systems flown in the center to the 3rd Space Experimentation Squadron in a realignment of functions. This realignment solidified 1 SOPS' functions as the wing's space situational awareness activity and placed all pathfinder missions under the 3rd Space Experimentation Squadron.

For much of the installation's history, medical services were provided by a small cadre of personnel assigned to Peterson AFB. At first, a few medical and dental technicians commuted to Falcon to provide "sick call" services, referring things more serious than colds or flu and routine dental exams to appointments with the primary care facility at Peterson AFB. As the Falcon workforce continued to grow, the small clinic took on a greater role in providing for the daily needs of the base's personnel, focusing on the active-duty force. Even as Falcon AFB neared its fifteenth anniversary, however, the clinic did not have a permanent, dedicated medical facility. Construction on a new, dedicated medical clinic began on July 1, 2003, and the facility opened on July 22, 2004. On October 1, 2003, the command activated a detachment of the 21st Medical Group at Schriever to provide full service dental and medical clinic services. Then, on June 22, 2012, Air Force Space Command activated the 21st Medical Squadron, a component of the Peterson AFB 21st Medical Group, to serve the Schriever community, signifying the growth of the detachment's responsibilities. Since that time, the squadron has expanded its services to the community, incorporating a mental health clinic as well as other services and expanding its clientele to include family members of military personnel and local military retirees.

Today, Schriever AFB hosts twenty-one tenant organizations and activities, while the 50th Space Wing operates or supports approximately 175 satellites. The base's proximity to key space organizations, including Air Force Space Command headquarters, the Cheyenne Mountain Complex, and the Colorado Front Range's growing space industry, were important factors in the base's growth over the past thirty years and they remain important today.

Reflecting this, the Department of Defense announced to the U.S. Congress on September 10, 2015, that Schriever AFB would be home to the experimental Joint Interagency Combined Space Operations Center (JICSpOC). With $16 million in fiscal year 2015 funding from the DoD and additional funding from the Director of National Intelligence, DoD leaders expected experimentation and testing to begin in October 2015 with integration of the results of those evaluations and the development of standard operating procedures by January 2017. The new agency was expected to employ thirty or more personnel and support the Joint Space Operations Center at Vandenberg AFB, California, as well as U.S. Strategic command's Joint Functional Component Commander for Space. The JICSpOC's initial role will be to improve data sharing between the intelligence community and military organizations, and to assist in countering adversarial threats to U.S. satellites and its other space-based systems.

Falcon
FEEDBACK

Volume 2, Number 9 | Falcon Air Force Base, Colorado | August 15, 1990

UK head of state speaks at Falcon

Photo by Sgt. Jon Hunt

British Prime Minister Margaret Thatcher addressed a gathering of over 400 Falcon people in the National Test Bed on Falcon AFB Aug. 3. Thatcher discussed the importance of the Strategic Defense Initiative as a deterrent to Soviet aggression after spending nearly three hours of touring and briefings in the NTB. Thatcher is the second head of state to visit Falcon AFB.

INSIDE:

New bosses Page 2

Levitow today Page 4

In the dorms Page 5

British Prime Minister Margaret Thatcher visited the Joint National Test Facility on Falcon Air Force Base during her visit on August 3, 1990. While visiting the installation, Prime Minister Thatcher addressed an audience of over 400 employees of Falcon AFB, discussing the importance of the Strategic Defense Initiative as a deterrent to Soviet aggression.

"I firmly believe that it was the determination to embark upon that SDI program and to continue with it that eventually convinced the Soviet Union that they could never, never, never achieve their aim by military might because they would never succeed." … British Prime Minister Margaret Thatcher, 3 August 1990, Falcon AFB, Colorado.

The Joint National Test Facility (later Ballistic Missile Defense Organization and then Missile Defense Agency) was created to support research and development, system-level tests and evaluation, and to provide operational and training support to U.S. combatant commands. Missile Defense Agency operations for missile defense capabilities have been executed at Schriever AFB since 1988.

Colonel Roger G. DeKok, commander of the 2nd Space Wing, and an unidentified officer cut the ribbon to open the operations center for Detachment 46, Air Force Technical Applications Center (AFTAC). AFTAC is charged with providing national authorities quality technical measurements to monitor nuclear treaty compliance and developing advanced proliferation monitoring technologies to preserve our nation's security. It is the sole organization in the federal government whose mission is to detect and report technical data from foreign nuclear explosions. AFTAC's nuclear event detection mission is directly linked to its nuclear treaty monitoring mission. AFTAC monitors signatory countries' compliance with the 1963 Limited Test Ban Treaty. This treaty prohibits nuclear testing anywhere but underground and prohibits the venting of nuclear debris or radiation from those tests into the atmosphere outside the country's national borders. AFTAC also monitors the Threshold Test Ban Treaty of 1974 and the Peaceful Nuclear Explosion Treaty of 1976. The 1974 treaty limits the size of underground nuclear tests to 150 kilotons, while the 1976 treaty prohibits the testing of nuclear devices outside of agreed treaty sites.

The 11th Space Warning Squadron, a component of the 460th Operations Group, 460th Space Wing, Buckley AFB Colorado was an important element of the Schriever AFB team. Activated at Falcon AFB in 1994, the unit inactivated in 2002 and was again activated in 2007. First charged with operating the ALERT system from its mission control center (pictured), the squadron later was assigned the task of operating the Space Based Infrared System (SBIRS) Highly Elliptical Orbit (HEO) from the Mission Control Station Backup ground station at Schriever. Unit personnel "provide U.S. and allied forces worldwide the next generation of missile warning, battlespace awareness, technical intelligence, and missile defense data through command and exploitation of Air Force Space Command's only taskable, space-based infrared, highly elliptical payload."

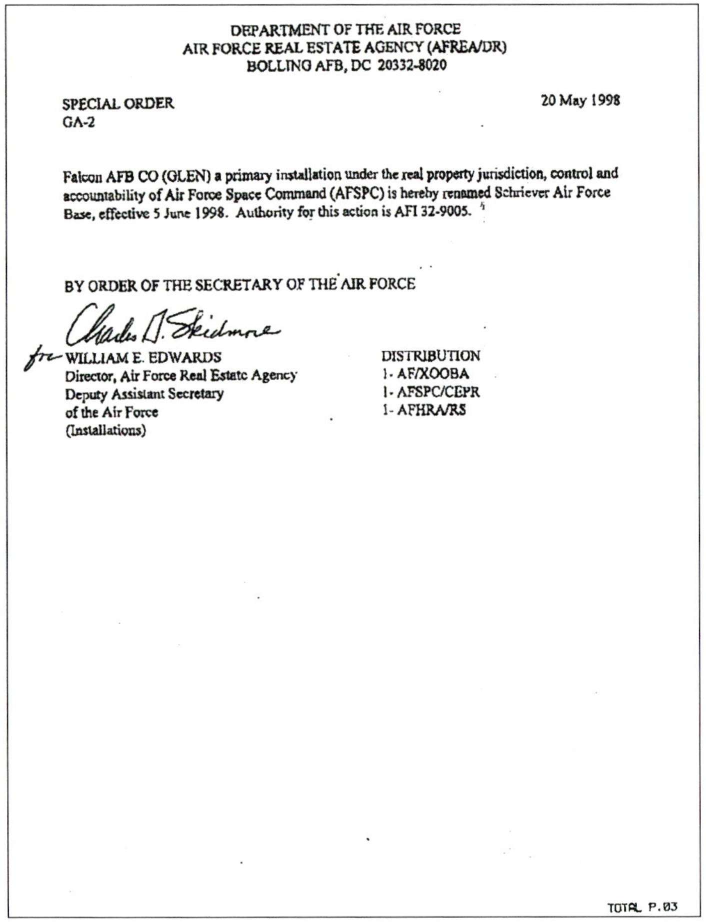

DEPARTMENT OF THE AIR FORCE
AIR FORCE REAL ESTATE AGENCY (AFREA/DR)
BOLLING AFB, DC 20332-8020

SPECIAL ORDER
GA-2

20 May 1998

Falcon AFB CO (GLEN) a primary installation under the real property jurisdiction, control and accountability of Air Force Space Command (AFSPC) is hereby renamed Schriever Air Force Base, effective 5 June 1998. Authority for this action is AFI 32-9005.

BY ORDER OF THE SECRETARY OF THE AIR FORCE

Charles A. Skidmore
for WILLIAM E. EDWARDS
Director, Air Force Real Estate Agency
Deputy Assistant Secretary
of the Air Force
(Installations)

DISTRIBUTION
1- AF/XOOBA
1- AFSPC/CEPR
1- AFHRA/RS

TOTAL P.03

On June 5, 1998, Headquarters, United States Air Force renamed Falcon Air Force Base in honor of General Bernard A. Schriever, the former commander of the Air Force's Western Development Center (for ballistic missile research) and Air Force Research and Development Command (later Air Force Systems Command). In 1997, Air Force Space Command elected him as one of the inaugural Space and Missile Pioneers.

One of the tasks of the Space Warfare Center was to monitor Year 2000 (Y2K) events. The common practice of truncating four-digit years to two digits presented concerns as the world approached the year 2000 not understanding how digital systems would respond to the change from 1999 to 2000 that would make the new year indistinguishable from 1900. The issue was raised in 1985 by Jerome and Marilyn Murray in their book Computers in Crisis (Petrocelli, 1984). Governments established special committees to monitor remedial work and prepare contingency plans, especially for critical systems such as communications, utilities, and others. Military systems received special emphasis. At the Space Warfare Center, United States Air Force and Russian officers supported Y2K testing to ensure that no issues arose threatening the strategic posture of forces around the globe.

Some of the changes at Schriever AFB resulted from efforts to incorporate new technologies and to update resources as funding for military manpower and contracting declined. The Remote Tracking Station (RTS) Block Change (RBC) program was one program affecting the 50th Space Wing's Air Force Satellite Control Network ground antennas that met these objectives. At Schriever AFB, construction of the new RBC antenna at the Colorado Tracking Station began on September 8, 2004 (top photo). Construction of the facility and installation of the radome were complete by July 2005. Additional budget constraints, however, resulted in the discontinuance of routine operations at the Colorado Tracking Station in July 2012. The site was subsequently decommissioned on September 29, 2014, following a brief period as a test site consisting of transportable assets. By summer 2015, the Colorado Tracking Station's facilities had been razed.

By 2005, there were 69 buildings on Schriever Air Force Base, which occupied more than 4,100 acres. The base's population reached 6,227 by early 2006. The base was renamed in honor of Gen. Bernard Schriever in 1998 and was the first base to be named for a living individual. In this May 2004 photo, the growth of the installation is readily apparent.

Personnel of the 1st Space Operations Squadron conduct their final support of the Defense Support Program satellite constellation by sending a power off command to Flight 15 on August 29, 2006, while Colonel Clinton E. Crozier, commander of the 50th Operations Group (second from right), and Lieutenant Colonel Stephen Mitchell (center) observe. The satellite previously had been placed in super-synchronous orbit during end-of-life operations. Although the squadron maintained the ability to conduct back-up operations, the 460th Space Wing at Buckley AFB, Colorado, assumed satellite command authority for the DSP constellation. The power-off command signaled the end of the squadron's 18-year association with the Defense Support Program. The squadron had conducted launch and early orbit operations, on-orbit command and control of the satellite bus, anomaly resolution, and end-of-life operations for the constellation since February 19, 1988.

Operational exercises played an important role in the ability of the base's personnel to fulfill their missions during emergency conditions. Exercise scenarios involved many different organizations and occasionally included components of the civil community's first responders. In the left photo, Schriever's emergency responders and an emergency medical technician transport a vehicle accident victim to a waiting flight for life helicopter during a casualty exercise. Above, Airman Robert Harris and an unidentified staff sergeant of the 50th Security Forces Squadron, search the Schriever Airman and Family Readiness Center for an active shooter during an exercise scenario. The exercise was conducted to ensure Schriever law enforcement and emergency services members are well trained and prepared.

From a "closet" to an operations floor. Until 1994, Detachment DELTA, Naval Satellite Operations Command (NAVSOC) functioned as the liaison between NAVSOC and the 3rd Space Operations Squadron (3 SOPS), the unit that provided telemetry, tracking and commanding (TT&C) for Navy communication satellites. At that time Detachment DELTA consisted of an officer-in-charge and an engineer. Additional engineers were later added for Ultra-High Frequency Follow-On (UFO) support. Mission requirements were changed in 2003 when the detachment was designated as the back-up command center for NAVSOC Headquarters after satellite control authority (SCA) for the Fleet Satellite Communications System (FLTSAT) was transferred to the Navy. The unit began to ramp up and, with the equipment installed and additional personnel assigned, became fully functional as NAVSOC's back-up satellite operations center.

Detachment DELTA first operated in what was known as the "closet", a room in Mod 18, owned by 3 SOPS. In 2004, the 50th Space Wing granted the use of Mod 19B to Detachment DELTA. In October of 2008 the process of combining Mods 19B and C to create a full operations floor was started. This quadrupled the size of the operations floor and facilitated updating aging equipment. The expansion also allowed space for Mobile User Objective System (MUOS) operations as well as launch and early orbit operations.

The Space Warfare Center (later renamed Space Innovation and Development Center) joined Team 5-0 on November 1, 1993 and remained a key partner with the 50th Space Wing until it inactivated on April 1, 2013. The center completed construction of its new facility in July 2009. One of the functions of the organization was to host the Schriever war games. The war game series began in 2001 to explore future space issues and integrate space capabilities throughout the United States national security infrastructure. North Atlantic Treaty Organization (NATO) and other allied partners have participated in the events over the 15 years. Responsibility for hosting the games transitioned to the USAF Warfare Center upon the inactivation of the SIDC. Pictured in the bottom image is the Senior Leadership Seminar for the 2010 Schriever Wargames, held at Nellis AFB, NV, in May 2010.

Following the inactivation of the SIDC, many of its units transferred to the USAF Warfare Center or were inactivated, while the 3rd Space Experimentation Squadron joined the 50th Operations Group.

Satellite crew personnel used the standard space trainer (SST) to conduct training in satellite command and control operations. The Schriever AFB SST facility opened in February 2011. Before the SST was available, wing crew personnel conducted satellite training on a variety of different systems. These legacy training systems varied not only in their look and feel, but in their structure. Varying hardware, operating systems, and proprietary software provided no clear training standard. The legacy systems left little room for versatility. Additionally, differences occurred in the systems used by Air Education and Training Command and Air Force Space Command for the same satellite system.

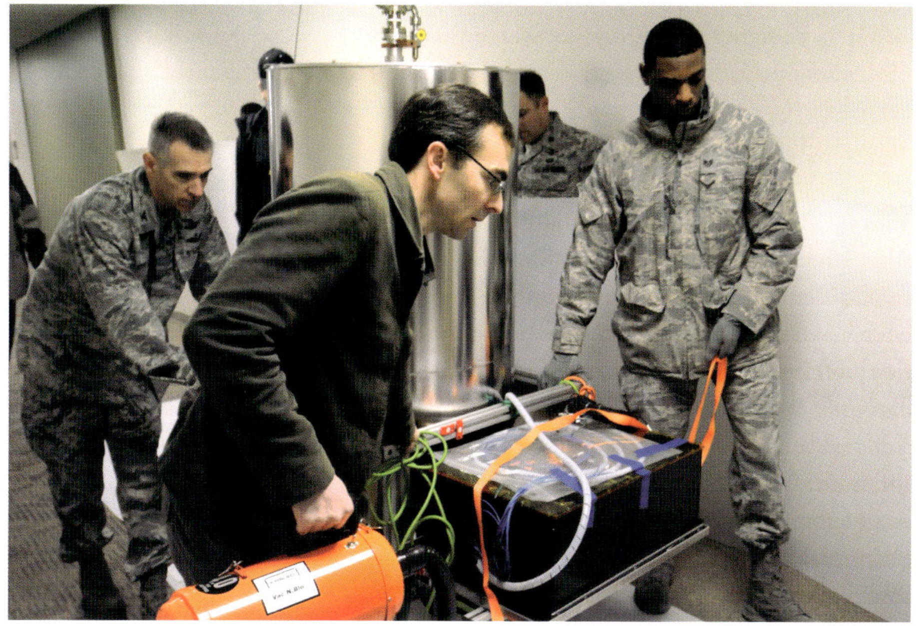

United States Naval Observatory physicist Dr. Steve Peil leads Schriever's new rubidium fountain clock to the master clock vault. In order to minimize damage to the equipment the clock was floated, much like a hovercraft, across a series of mats until it reached its final destination. The United States Naval Observatory first placed an alternate master clock at Schriever AFB in November 1995 to be co-located with the Global Positioning System.

The 50th Space Wing gained responsibility for the Wideband Global Satellite Communications (WGS) System and Advanced Extremely High Frequency (AEHF) satellite system as follow-on systems for the Defense Satellite Communications System and Milstar system, respectively. Each of the new satellites provided nearly ten times the capability of the legacy systems they would ultimately replace.

Prior to these systems becoming operational, the wing began planning for an integrated operating area that would streamline communications between the squadrons, between the units and wing leadership, and provide a clearer focus on operational concerns and events. Funding constraints winnowed the wing's vision to a combined operating environment for military satellite communications systems to be jointly occupied by the 3rd and 4th Space Operations Squadrons. The wing's Integrated Operations Environment (IOE) opened in January 2013. Today, the IOE is the home of the newly formed Joint Interagency Combined Space Operations Center (JICSpOC).

Major Shanna Corbet, 19th Space Operations Squadron reviews transfer procedures with Airman First Class Alan Faeldan, 2nd Space Operations Squadron during the Civil Navigation implementation April 28, 2014, in the 2 SOPS operations environments. The initiative, announced by the DoD will allow Air Force Space Command to broadcast Civil Navigation messages on all operational GPS satellites capable of transmitting the L2C and L5 signals.

The 310th Space Wing's 19th Space Operations Squadron (Air Force Reserve Command) operates the GPS system jointly with the 50th Space Wing's 2d Space Operations Squadron. Similarly, the 310th Space Wing's 6th and 7th Space Operations Squadrons support other 50th Space Wing satellite operations functions.

U.S. Navy Adm. Cecil D. Haney (right), U.S. Strategic Command commander, and senior Department of Defense (DoD) leaders, including Gen. David L. Goldfein (center), U.S. Air Force vice chief of staff, and Ms. Stephanie O'Sullivan (fourth from right), Principal Deputy Director of National Intelligence, receive a briefing from General John E. Hyten, Air Force Space Command commander, while attending the Joint Interagency Combined Space Operations Center's (JICSpOC) distinguished visitor day at Schriever AFB. The JICSpOC, a collaborative effort between USSTRATCOM, AFSPC, the NRO, and the intelligence community and commercial data providers, began operational testing in October 2015.

Col. DeAnna Burt, 50th Space Wing commander, briefs the 50 SW mission during the State of the Base Wednesday, March 2, 2016, at Schriever Air Force Base, Colorado. State of the Base provides an opportunity for Schriever leaders to strengthen relationships with Colorado Springs and El Paso County civic and community leaders.

Having grown from an operations-focused station of 640 acres and a couple of hundred personnel, as of May of 2014, Schriever AFB encompassed 3,840 acres holding 63 buildings (not including housing units) and reported a population of 7,647 personnel. Managing the infrastructure to support the installation, its employees, and families was principally the responsibility of the men and women of the 50th Mission Support Group, a component of the host 50th Space Wing.

Mid-way through the first decade of the 21st century, Schriever AFB hosted nearly 60 major and minor facilities and employed over 6,200 people. The base's continuing growth and importance prompted wing and command officials to begin preliminary planning to bring several hundred housing units and associated community support activities to the base. That effort came to fruition when, on May 16, 2008, base officials, community leaders, and contractor representatives broke ground on 242 enlisted and officer housing units at Schriever AFB. Construction was completed by late summer 2010. Today, 223 families call Schriever AFB home.

It wasn't the construction of base housing, however, that built the Schriever community. Instead, the men and women--civilian, contractor, and military personnel--that operated Schriever's space-based systems, managed the infrastructure, and supported combatant forces worldwide adopted the installation as their home town and provided their services, expertise, and time to the local communities around the base, including Falcon and Ellicott, Colorado. And, the communities reciprocated.

Early in the base's history, the installation's first responders—security forces and firefighters—inked mutual aid agreements with civil agencies. In thirty years, the base's security forces and firefighters have responded to a number of traffic accidents, wild land, and grassland fires off the installation, providing emergency first aid and other capabilities. These examples, perhaps, were best illustrated in the base's response to two wild land fires in the Colorado Springs region.

The Waldo Canyon fire in western El Paso County, Colorado, began on June 23, 2012, at approximately 12 p.m. in moderately to extremely difficult terrain. The blaze, which lasted approximately 17 days, consumed over 18,000 acres, 346 homes, and claimed two lives making it arguably the most destructive fire in Colorado history. During the incident, local military installations responded to provide temporary shelter for evacuees, assist with fire-fighting operations, and provide donations of needed blankets, clothing, and food for those affected by the fire.

On June 26, 2012, Colonel James P. Ross, responding to civil requests for resources to fight the fire in Waldo Canyon, west of Colorado Springs, authorized one engine, four firefighters, and one fire safety officer to deploy to the United States Air Force Academy (USAFA) to assist with efforts to fight the fire as it reached USAFA property. On-scene crews were able to contain the flare-up to approximately 10 acres. Additionally, a SAFB firefighter joined the El Paso County Emergency Operations Call Center, per the base's mutual aid agreement with the county, as did a member of the wing's Public Affairs staff.

Wing personnel also assisted with the post-fire activity and volunteered to support the community in other ways. Thea Wasche, deputy commander of the 50th Force Support Squadron and her therapy golden retriever, Lacey, visited evacuation shelters to bring comfort to those displaced. After the fire, members of the 1st Space Operations Squadron, Technical Sergeants Jason Childers and Gaston Lara, and Captains Dan Coleman and Stan Maczek assisted with clean-up operations. Other Schriever AFB personnel and organizations collected food and other necessities to assist displaced persons or contributed in countless other ways.

Eleven months later, a wild land fire erupted northwest of Schriever AFB in the Black Forest area of El Paso County. The conflagration, which began on June 11, 2013, quickly moved north and east into residential areas, pushed by gusty winds and dry timber. On June 12, 2013, civilian authorities asked the base for assistance and the 50th Civil Engineer Squadron dispatched an engine and crew that eventually helped make up a military response team comprised of assets from Fort Carson, Peterson AFB, Schriever AFB, the Air Force Academy, and the Colorado National Guard. Schriever's fire department continued to deploy crews around the clock until relieved on June 16, 2013, at which time the fire was approximately seventy-five percent contained.

Once the fire was extinguished, Schriever personnel repeated their efforts following the previous year's Waldo Canyon fire by assisting with clean-up operations, collecting and distributing food, clothing, and other essentials for displaced homeowners, and provided other support. This fire proved especially distressing to the Schriever community as the two fatalities resulting from the fire had worked at the base for many years.

On May 28, 2001, a destructive tornado ripped through Ellicott, damaging homes and the high school under construction and scheduled to open in the fall. As in later emergencies, base personnel responded with collections of food and other essentials for those affected by the tornado. Schriever AFB personnel also participated in clean-up operations in the days and weeks following the tornado, saving residents some of the burdensome cost of removing large debris and materials.

It is a common practice for military organizations to "adopt" local roadways and periodically clean up litter and debris along those roadways. Units at Schriever AFB were not the exception. Many of the base's organizations adopted portions of local streets and roads. Base personnel also contributed time and effort to many local charities, in addition to the annual Combined Federal Campaign, in which Schriever set a standard of meeting or exceeding is goal in most years. Schriever's employees contributed to Habitat for Humanity in its efforts to build homes for local disadvantaged families. They supported local Boy Scout and Girl Scout troops and other youth organizations. Schriever's personnel also supported local schools, providing mentoring and other support to students and the districts. Local Junior Reserve Officer Training Corps (JROTC) squadrons toured the base and learned of its satellite and related missions, visiting 50th Space Wing satellite operations centers and other important facilities and units. Schriever AFB personnel volunteered to mentor to local students, working to improve reading skills and improve scholastic performance.

Schriever-based units joined other local military installations providing formations and floats for holiday parades. They participated in the annual Pikes Peak or Bust Rodeo and other events. Schriever commanders met regularly with local civic leaders, including the city's Military Affairs Committee and through the quarterly Chief of Staff luncheon. Maintaining these relationships with local officials were important aspects of the commanders' priorities. The wing's Honorary Commanders Program was designed to increase public awareness of Schriever AFB and its missions, people, and programs and to build relationships with the local community, encouraging them to share the base's story. In 2012, the 50th Space Wing revitalized the program and inducted seven honorary commanders. These seven honorary commanders were granted alumni status two years later and eight new honorary commanders were introduced to wing, group, and some squadron leaders in 2014.

And the communities of the Front Range reciprocated. Sports teams, local attractions, and businesses supported the base and its population by sponsoring events or providing discounted tickets to their events. That support was evident in the level of commercial sponsorship provided to the installation to support, for example, the base's annual summer picnic. Through local sponsorship, the base was able to fund a variety of events, meals, and competitive prizes. Local sports teams and attractions made discounted (and occasionally free) tickets available to employees at Schriever AFB, as well as the other military installations on the front range.

Schriever's community involvement was certainly not limited to interactions with the local civilian populations; although that interaction was important and base officials worked to maintain close relationships with local officials. Equally important, the base's supporting organizations, especially the agencies of the 50th Force Support Squadron and the base Chapel, developed and instituted programs to maintain the health and wellness of the base population. From operating fitness facilities and programs to special activities for single Airmen, families, or married couples, these initiatives and programs aided in maintaining a sense of community and well-being in a fluid population.

The construction of housing on Schriever AFB furthered the development of the Schriever community and the installation's support to local schools as dependents living on the installation attended classes in the Ellicott School District. The 242 housing units new on Schriever AFB in 2010 brought about 740 family members to live on the installation. Tierra Vista Communities, the development and management company operating Schriever's housing complex, initiated programs designed for residents and supported other units' efforts to foster a sense of community across the base. The natural result was an increased focus and partnering with the rural district. School and installation officials worked together to promote studies in science, technology, engineering and mathematics (STEM), and Ellicott school children were often invited to participate in events held at the base, including the annual holiday tree lighting, "trick-or-treat" night, and others. Base agencies visited Ellicott schools, providing information on fire safety, drug-free living, and similar programs.

Building a strong community at Schriever AFB meant providing essential services and infrastructure to the base. In the early 2000s, the base underwent a dramatic change as facilities to provide essential services were constructed. Between 2001 and 2005, contractors completed construction projects totaling more than $58 million. New facilities included a community support center (2002), fitness center (2001), medical and dental clinic (2004), an addition to the 50th Space Wing headquarters building (2004), and some operational facilities including a backup operations center for the Space Based Infrared System (2003). The base's Secure Area Logistics Facility and a new west gate also opened in 2004.

As the base workforce grew, leaders knew that additional community support infrastructure was necessary. Construction for the base's Child Development Center, a key component of the base's long-range planning, began with a ground-breaking ceremony on February 14, 2001. Eighteen months later, on September 6, 2002, the facility opened. By May 2016, the Child Development Center provided care for approximately 210 children of Schriever AFB employees.

Twenty-five years after the ground-breaking for the installation, Colonel Theresa A. Djuric, commander of the 50th Space Wing, joined by representatives of Tierra Vista Communities and other members of the Schriever AFB broke ground on the base's 242-unit housing development on May 16, 2008. By August 2009, the first units were available for occupancy and Second Lieutenant Caleb Murphy and his wife, Christina, became the first occupants. By August 2010, Tierra Vista Communities, a subsidiary of Actus Lend Lease, had completed construction of all units and the community center by March 22, 2010.

The effort to bring enlisted and officer housing to Schriever AFB had been long. Base and command officials lobbied Air Force leadership and ultimately the United States Congress to fund housing for the installation as a component of a three-installation program that included new housing construction at nearby Peterson AFB and at Los Angeles AFB in California.

Schriever AFB personnel engage in science experiments with Ellicott School District children October 8, 2015. The installation maintained a close relationship with the school and participated in a number of educational and public service awareness campaigns with the students. The science experiments were part of a public program to expose elementary students to science, technology, engineering, and math (STEM) studies. The base also supported fire prevention and awareness week and Red Ribbon week activities designed to make students aware of the dangers of drug use.

One of the elements of the base's efforts to build a strong community was maintaining and improving the natural environment around the base. For the seventeenth consecutive year, in 2015 Schriever AFB was named a "Tree City USA" by the Arbor Day Foundation. Andy Schlosberg, Colorado State Forest Service Assistant District Forester, Marc Williams, Col. DeAnna Burt, 50th Space Wing commander, Col. Brian Barthel, 50th Mission Support Group commander, Lauren Ouding, and Lt. Col. Nathan Clemmer, 50th Civil Engineer Squadron commander, help plant a tree during the Arbor Day celebration at the Child Development Center. The tree planting during Arbor Day was an annual event at Schriever Air Force Base.

Ellicott Schools played an important role in the Schriever AFB community. School-age children residing in the base's housing community attended Ellicott schools. Each year, the Ellicott Elementary School Choir visited the base for the holiday tree lighting ceremony and entertained base personnel with a variety of holiday songs. Following the performance and the lighting of the tree, the children and others enjoyed a visit with Santa Claus, hot chocolate, and cookies in the 50th Space Wing headquarters building.

Schriever Air Force Base's community involvement exceeded the front range and the state of Colorado. Schriever's personnel served as unofficial ambassadors of the installation and the Air Force around the world. Capt. Colin Merrin, 2nd Space Operations Squadron, Capt. Heidi Kent, 7th Space Operations Squadron, and Capt. Megan Harkins, 1st Space Operations Squadron, pause to represent the 50th Space Wing at Mount Everest Base Camp. Capt. Merrin, a member of the USAF 7 summits team, had to abandon his ascent at 27,000 feet on May 21, 2013, due to health concerns.

From left: Tech. Sgts. Jason Childers and Gaston Lara and Capts. Dan Coleman and Stan Maczek pose for a photo with a contractor foreman (middle) July 26, 2012, while volunteering their time to help clean up the Waldo Canyon burn site. The Airmen are all members of the 1st Space Operations Squadron. (Courtesy photo)

Schriever firefighter Scott Bartlett extinguishes a hot spot while helping fight the Black Forest wildfire June 14. The Black Forest fire, one of the most destructive (in terms of dollar value of damaged properties) in Colorado history began in hot, windy, dry conditions only one year after the Waldo Canyon fire.

Schriever AFB's location about nine miles east of Peterson AFB meant that typical recreation venues for assigned personnel, such as bowling alleys, clubs, or other amenities were not available. To compensate for the lack of these types of programs, the 50th Force Support Squadron's Services Flight offered special programs for single Airmen under the Single Airman Programming Initiative (SAPI). Schriever began its participation in the initiative in fiscal year 2012, and since then involved 739 persons in 71 events from skydiving to fly fishing, snowmobiling, and pheasant hunting. The purpose of the SAPI was to create a culture supporting the "whole-person" concept.

Personnel of the 4th Space Operations Squadron and the 50th Operations Group, Capt. Rodrigo Ocampo (orange hardhat), 1st Lt. Mark Long (behind ladder), 2nd Lt. Aaron Blore (green hardhat), and Capt. Aaron Doyle (blue hardhat), erect a ladder to facilitate roof truss installation during a Pikes Peak Habitat For Humanity construction project in Colorado Springs. Schriever's employees have, throughout the base's history, contributed to many local charitable efforts. Lieutenant Long set up the base's Habitat For Humanity volunteer crew monthly coordinating the efforts of about 40 volunteers. (Courtesy photo)

SCHRIEVER AFB COMMANDING OFFICERS

2nd Space Wing

Colonel Richard L. Griffin .... July 8, 1985–July 7, 1986
Brigadier General Lester Weber .... July 8, 1986–December 11, 1988
Brigadier General Jimmey Morrell .... December 12, 1988–August 26, 1990
Colonel Roger DeKok .... August 27, 1990–January 30, 1992

50th Space Wing

Brigadier General Roger G. DeKok .... January 30, 1992–June 16, 1993
Colonel Gregory Giles .... June 17, 1993–November 3, 1994
Colonel Simon P. Worden .... November 4, 1994–March 21, 1996
Brigadier General Glen W. Moorhead III .... March 22, 1996–April 24, 1997
Colonel Elwood C. Tircuit .... April 25, 1997–June 9, 1999
Colonel Richard E. Webber .... June 9, 1999–April 19, 2001
Colonel Larry D. James .... April 20, 2001–June 8, 2003
Colonel Michael D. Selva (Interim) .... February 7, 2003–May 2003
Colonel Suzanne M. Vautrinot .... June 9, 2003–April 3, 2005
Colonel John E. Hyten .... April 4, 2005–May 21, 2007
Colonel James C. Hutto, Jr. (Interim) .... May 15, 2006–October 13, 2006
Colonel Teresa A.H. Djuric .... May 22, 2007–June 11, 2008
Colonel Cary C. Chun .... June 12, 2008–August 19, 2009
Colonel Wayne R. Monteith .... August 20, 2009–August 4, 2011
Colonel James P. Ross .... August 5, 2011–July 10, 2013
Colonel William J. Liquori, Jr. .... July 11, 2013–May 28, 2015
Colonel DeAnna M. Burt .... May 29, 2015–

11th Space Warning Squadron: The squadron is a geographically-separated unit of the 460th Space Wing at Buckley AFB, Colorado. The mission of 11 SWS is to provide U.S. and allied forces with worldwide missile warning, battle-space awareness, technical intelligence and missile defense data through space-based infrared, highly elliptical operations.

21st Medical Squadron: The mission of the squadron is to provide seamless, patient-focused care to the Schriever beneficiary population. This includes diagnostic, preventative, and treatment related health care services. The squadron also provides patient education and continuity of care for health maintenance as well as for the acute and chronic management of disease and injury.

53rd Signal Battalion (Satellite Control): The mission of U.S. Army Strategic Command's 53rd Signal Battalion is to operate, maintain and sustain three contingency Defense Satellite Communications System control platforms to provide positive platform, payload and communications network control of DSCS and Wideband Global SATCOM satellites to the unified and specified combatant command commander, the National Command Authority and other customers approved by the Joint Staff.

100th Missile Defense Brigade, Missile Defense Element: The 100th Missile Defense Brigade's Missile Defense Element is located in the MDIOC. MDE functions include conducting the operational level execution of the Ground Based Midcourse Missile Defense Limited Defensive Operations mission for the U.S. Northern Command commander. In this respect, MDE provides operational and tactical recommendation to the NORTHCOM commander, synchronizes operations and conducts fire distribution across multiple GMD Fire Direction Centers, conducts battle planning for potential follow-on launches of threat ballistic missiles and, as required, de-conflicts the operational pictures for warning and missile defense missions. MDE also manages Army GMD force readiness conditions and develops and reports overall GMD operational capability. It provides day-to-day control of GMD FDCs, performs analyses and develops responses to NORTHCOM commander, command director and missile defense officer questions and provides back up mission execution capability for the firing unit FDCs.

310th Space Wing: The 310th Space Wing, the only space wing in the Air Force Reserve, has units assigned to Schriever, Peterson and Buckley Air Force Bases in Colorado, as well as Vandenberg Air Force Base in California. The wing provides optimized, scalable combat ready forces delivering space and cyberspace power for faster, more lethal, more accurate affects every day, everywhere.

527th Space Aggressor Squadron: The squadron prepares USAF, joint and allied forces for combat through realistic threat replication, training, and feedback through specialized and certified space-capable aggressors. It operates adversary space systems, develops new tactics, techniques and procedures to counter threats, and improves the US military space posture.

Air Force Technical Applications Center, Detachment 46: Air Force Technical Applications Center, Detachment 46 supports the Nuclear Detonation Detection System payload on board the Global Positioning System satellite constellation. NDS data is used in AFTAC's worldwide Nuclear Test Ban Treaty monitoring system, the Integrated Tactical Warning and Attack.

Joint Functional Component Command for Integrated Missile Defense: The headquarters is located in the Missile Defense Integration-Operations Center at Schriever Air Force Base, Colorado. The BMDS is operational to defend against all ranges of ballistic missile threats. The command's location allows JFCC IMD to leverage the existing infrastructure and its strong partnerships to execute IMD planning and operational support responsibilities. This modern defense network is expanding to include radars and other sensors that feed information from the air, land, sea, and space environments to battle management centers that can direct interceptors to targets. The JFCC IMD began operations in February 2005.

Missile Defense Agency: The Missile Defense Agency's (MDA) mission is to develop, test, and field an integrated, layered, ballistic missile defense system (BMDS) to defend the United States, its deployed forces, allies, and friends against all ranges of enemy ballistic missiles in all phases of flight.

Missile Defense Integration & Operations Center: The Missile Defense Agency's Integration and Operations Center supports research and development, system-level tests and evaluation, and provides operational and training support to U.S. combatant commands. Missile Defense Agency operations for missile defense capabilities have been executed at Schriever AFB since 1988.

Naval Satellite Operations Center Detachment Delta: The Naval Satellite Operations Center Detachment Delta is located in building 400. The unit's primary mission is to provide hot backup operations for NAVSOC headquarters located in Point Mugu, Calif. Primary constellations supported include Fleet Satellite, Ultra High Frequency Follow On, and Mobile User Objective System (MOUS).

Space and Missile Systems Center, Detachment 11: The Space and Missile Systems Center, a subordinate unit of Air Force Space Command, is the center of technical excellence for developing, acquiring, fielding and sustaining military space systems. SMC's mission is to deliver resilient and affordable space capabilities. The center is responsible for on-orbit check-out, testing, sustainment and maintenance of military satellite constellations and other Department of Defense space systems.

United States Air Force Warfare Center: The USAFWC exists to ensure deployed forces are well trained and well equipped to conduct integrated combat operations. From Testing and tactics development programs to training schools and venues, the USAFWC provides Airmen. The USAFWC vision, mission and priorities are central to supporting the Air Combat Command's mission to provide dominant combat airpower for America with warrior Airmen committed to excellence, trained to fly, fight, and win...anytime, anyplace. Several units of the USAFWC, including the Air Force Tactical Exploitation of National Capabilities (AFTENCAP), operate from Schriever AFB.

United States Naval Observatory, Alternate Master Clock: The United States Naval Observatory provides a wide range of astronomical data and products. USNO products support activities in astrometry, astronomical applications, Earth orientation, and precise time. The USNO serves as the official source of time for the U.S. Department of Defense and a standard of time for the entire United States. This segment of the USNO mission enhances many applications, including space operations, command, control, communications, computer networks, intelligence, surveillance, and reconnaissance (C4ISR), positioning, navigation, and timing (PNT). The Alternate Master Clock is the backup of the USNO's Master Clock facility in Washington, D.C., and provides precise time to the 50th Space Wing and tenant organizations on Schriever AFB.